DEVELOPMENTAL AND MODERN BIOLOGY

DEVELOPMENTAL
AND
MODERN BIOLOGY

Editor
G.C. Satpathy

KALPAZ PUBLICATIONS
DELHI-110052

DEVELOPMENTAL AND MODERN BIOLOGY

(Biology)

ISBN: 81-7835-053-X

Published in 2001 in India by
Kalpaz Publications
C-30, Satyawati Nagar
Ashok Vihar Phase-III
Delhi-110052
E-Mail: kalpaz@hotmail.com

Laser Typesetting at : A.S. Printers
Printed at : Mehra Offset Printers, New Delhi

Preface

Biology is the science of the structure and function of plants and animals. It is often divided into botany—the science of plants, and zoology—the science of animals. More than hundred years ago, botanists and zoologists alike believed that many of the life processes of plants and animals would be discovered when life was observed under a microscope. By examining a wide variety of living organisms, the conclusion was reached that there was a substance common to all; the living matter appeared as a translucent, greyish material, usually capable of flowing freely. As every subject in biology, and in science generally, developmental biology also bears in it the excitement of discovery and of the birth and growth of ideas.

Developmental and Modern Biology brings information on the latest developments and critical analysis of data on a wide and useful range of specialised areas of biology. Fundamental principles are used as frameworks for presenting specific information, and research trends are included to provide a glimpse of future developments in the field, tried to cover enough areas to give a sense of unity and diversity to modern developmental biology.

This book provides a comprehensive survey of major issues at the forefront of modern developmental biology.

These information should be valuable to researchers in the fields of animal and plant development, as well as to students and professionals who want an introduction to current topics in cellular and molecular approaches to developmental biology.

Editor

Contents

1.	Introduction to Developmental Biology	1
2.	Evolutionary Biology	49
3.	Embryology and Genetics	74
4.	Sex Determination	103
5.	Gene Expression	124
6.	Homeobox and *pax* Genes	161
7.	Biological Membranes	191
8.	Petal and Stamen Development	211
9.	DNA Technology and Ecological Evolution	235

Chapter 1

Introduction to Developmental Biology

The only way to glean some understanding of evolutionary change is to persist in the historical approach of interpreting comparative studies of development in light of phylogenetic relationships of the species being compared. From this work it might be possible to draw inferences about the developmental processes of the ancestral species, and hence about the developmental changes that have occurred during evolution. In using comparative development to understand evolutionary change, it is important to Consider how we arrive at the phylogenetic tree on which the comparisons are based. Obviously, the conclusions drawn from comparative studies hinge on our assumptions about the phylogenetic relationships of the species being compared. Unfortunately, to the extent that embryological characteristics have been used in constructing phylogenetic trees, the use of these same phylogenetic trees as a framework for comparative developmental studies can lead to circular reasoning. Although DNA sequence comparisons per se are of little help in understanding the evolution of development, they do provide a way of constructing phylogenetic trees independent of data from comparative morphology or comparative embryology. This means that, for the first time, the problem of constructing phylogenetic trees can be divorced from the problem of understanding the developmental changes by which the trees have arisen. What sort of results can be expected from such comparative studies? One result will certainly be the identification of developmental processes that tire widely seen in modern species and are therefore apt to have been inherited more or less intact from the ancestral species.

Since genetic mutations are a fundamental component of evolutionary change, one might propose to study phyletic change simply by comparing DNA sequences from the species of interest. But this approach is unworkable because knowledge of genomic sequences is insufficient to explain developmental processes. This failure might simply reflect our present, rudimentary understanding of the cellular and molecular basis of embryogenesis or it might reflect a theoretical barrier to predicting the properties of complex systems from the properties of their simple components. One aspect of leech development that we seek to understand is how the rostrocaudal patterning of the body plan is generated. In addressing this issue, it is important to distinguish between "segmentation" and "regionalization" as two distinct features of rostrocaudal patterning. Segmentation refers to the organization of the body into a series of similar units, or segments, along the rostrocaudal axis. Regionalization refers to the division of the animal into dissimilar regions (e.g., head, trunk, and thorax) along its length. In segmented animals, regionalization is reflected in the fact that the segments are similar, but not identical, to one another along the rostrocaudal axis. The importance of distinguishing between segmentation and regionalization comes from the fact that these processes probably arose independently during evolution and therefore might operate independently during development. Regionalization in animal body plans had clearly evolved long before the separation of the lineages leading to protostomes and deuterostomes: regionalization is obvious in various pseudocoelomate phyla that give no indication of segmentation.

Comparing the segmentation processes in annelids and other segmented protostomes is particularly interesting because these groups are generally considered as having a common ancestor that was already segmented. Thus, it is expected that they share a common molecular-genetic basis for segmentation, as well as for regionalization. And yet the cellular processes of rostrocaudal pattern formation in protostomes vary dramatically, with long germ band insects and leeches at two extremes.

A priori, one can imagine segments arising either Simultaneously throughout the length of the embryo from a preexisting field of cells, or one by one in a strict progression starting at one end of the embryo. For convenience, these two logical extremes will be referred to as the "spatial" and "temporal" modes of segmentation, respectively. Most readers will probably recognize immediately that the spatial mode of segmentation is a good approximation of the course of events in the *Drosophila* embryo and other long germ band insects. Less widely appreciated, however, is the fact that the temporal mode, commonly associated with chordate development, is also observed in protostomes.

Homologous developmental processes should exhibit homologies at the molecular biological and genetic levels. This is the reasonable starting assumption for comparative developmental studies. Thus, the extensive molecular-genetic studies of segmentation in *Drosophila* should provide a solid foundation for analysing the process in leech, for which no genetic techniques are presently available. We are pursuing the now common procedure of identifying and characterizing homologous in leech of genes that have been identified as developmentally important in other organisms, especially *Drosophila*.

Identified sets of interacting selector genes have been shown to play roles in specific aspects of early development of the *Drosophila* embryo. Moreover, the groups of selector genes involved in one aspect of development may operate quite independently of those involved in other aspects.

For example, the groups of genes involved in establishing the dorsoventral patterning of the *Drosophila* embryo operate independently of those involved in rostrocaudal patterning. Among the genes involved in rostrocaudal patterning, as described earlier, those late-acting genes involved in regionalization operate quasi-independently, of late-acting genes involved in segmentation. Such sets of cross-regulating selector genes have been called syntagmata. Just which genes should be included in a particular syntagma in a given species is difficult to establish because the strengths of interactions between genes vary and cannot be experimentally determined with precision. But a corollary of Garcia-Bellido is definition of the syntagma is that such interacting genes and their operational role in development should be transmitted and modified collectively rather than individually during evolution.

The term syntagmal homology will be used to refer to the conservation of developmentally significant gene function between species in addressing the question of what is preserved as developmental processes change during evolution. Sequence homologs of many of the developmental regulatory genes from *Drosophila* have been identified outside of the arthropods. Insofar as different sequence homologs have been demonstrated to encode transcription factors, they can be classified as biochemical homologs as well. In addition the homologs can be compared at the level of spatial and temporal expression patterns. For example, one can ask whether they are expressed in homologous tissues and at comparable times in development. In keeping with the evolutionary antiquity of the regionalization process, homologs of the homeotic gene complexes, designated as the HOM/Hox complex, appear to be involved in the regionalization of animals ranging from nematodes to vertebrates. In light of this, it is not at all surprising that HOM/Hox genes appear to be involved in annelids as and it seems likely that these genes are involved in the regionalization process of leeches.

The tubular body is organized into 32 segments. Most of its length is occupied by 21 midbody segments, numbered M1-M21. In addition, there are four fused rostral segments (R1-R4) and seven fused caudal segments (C1-C7). Segments are strikingly similar to one another, especially within the midbody: for instance, each segmental ganglion consists of stereotyped numbers of individually identified neurons most of which have obvious homologs in other ganglia along the length of the animal. Despite this similarity between segments, segment-specific differences in both neural and nonneural tissues provide evidence of regionalization of the leech body plan. This, plus the fact that the number of segments is invariant, makes it clear that individual segments are assigned specific identities at some point in the development of leeches as in arthropods. The anterior end of the adult leech has nonsegmental tissues, such as a pharynx and a supraesophageal ganglion: collectively, these tissues are referred to as the prostomium. The gut and reproductive tissues are also generally regarded as nonsegmental tissues.

The embryonic development of glossiphoniid leeches is divided into 11 stages. The overt cellular processes leading morphologically recognizable segments occur during stages 6 to 9. In the species used in our research (primarily *Helobdella triserialis. H. robusta,* and *H. stagnalis*), eggs (each about 0.5 mm in diameter) are fertilized and internally, but arrest in first meiosis until deposited in transparent cocoons attached to the ventral surface of the parent: zygotes and embryos can be easily removed for experimental manipulation at any time during development. Stages 1 to 6 consist of a cleavages, giving rise to three distinct classes of individually identifiable blastomeres: macromeres, micromeres, and teloblasts. The 25 micromeres (3 arise from each of the A, B, and C quadrants of the embryo, and 16 arise from the D quadrant) contribute epithelical cells to a transient larval integument, as well as neurons, muscle cells, and epidermis to the prostomial tissues of the leech, whereas the 10 teloblasts generate the segmental tissues. Teloblasts are five bilateral pairs of embryonic stem cells that also arise from the D quadrant of the embryo. One pair of teloblasts, designated M, gives rise to the mesoderm; and four pairs of teloblasts, designated N, O/P, O/P, and Q, together give rise to the ectoderm, immediately overlying the mesoderm. Segmental organization of the ectoderm and mesoderm is first manifested in the direct descendants of the teloblasts. Beginning an hour after its birth (i.e. stage 5 for the M cells and stage 7 for the O/P cells), each teloblast repeatedly divides from a single anterior budding site, giving rise to much smaller primary blast cells. Each blast cell remains in contact with the one born immediately before it and with the one born immediately after it. Consequently, the blast cells are organized into coherent columns (bandlets) in the order of their individual birth rank. Ultimately, the first-born blast cells contribute their definitive progeny to anterior segments, and later-born cells contribute to the posterior segments. The five bandlets on each side of the embryo join together to form structures called germinal bands which lengthen as blast cells are produced at the posterior end of the bandlets.

The germinal bands on each side of the embryo meet at the site of the future head and coalesce progressively from anterior to posterior along the future ventral midline, forming a structure called germinal plate. In normal development, each teloblast contributes a distinct, segmentally iterated set of definitive progeny to the leech via its primary blast cell progeny.

The composition of these segmentally iterated sets of cells, called M, N, O, P, or Q kinship groups, can be predicted from the identity of the parent teloblast, except in the case of the O/P teloblasts; ipsilateral O/P teloblasts arise from an equal cleavage as Indistinguishable sister cells.

In fact, each O/P teloblast and its immediate progeny are developmentally plastic, capable of generating either O or P kinship groups. In any given *Helobdella* embryo, the particular fate of the O/P teloblasts can be determined retrospectively, at which point they can be designated as "generative O" and "generative P" teloblasts. During stage 9, segments become apparent in the germinal plate in an anterior-to-posterior progression, first through the delineation of mesodermal hemisomites and later through the appearance of segmental ganglia along the ventral midline. During stage 10, the edges of the expanding germinal plate meet along the dorsal midline, thereby closing the body tube. By the end of stage 11, the segmental tissues have differentiated to a state approximating their Mature form.

One attractive but short-lived proposal for homology between leech and fly development held that each teloblast underwent five rounds of syncytial nuclear divisions to generate multiples of 32 nuclei, which were then parcelled out by the subsequent blast cell divisions. This proposal had the advantage of providing a point of homology between annelids and arthropods, as well as a mechanism for counting out the correct number of segmental founder cells. Cell lineage analyses of segmentation Suggest other possible points of homology. One important set of observations concerns the relationship between primary blast cells and the final segments. In three of the segmental lineages (M, O, and P), a single primary blast cell makes one segment's worth of progeny, whereas two primary blast cells are required in the N and Q lineages.

As discussed in greater detail in the next section, the two n and two q blast cells that contribute to a segment can be assigned distinct identities on the basis of various differences. The first to be observed was a difference in the timing of their first mitoses; thus, they are referred to as nf (first-to-divide) and ns (second-to-divide) and qf and qs blast cells, respectively. A prominent feature of metamerism in *Drosophila* is the establishment of subdivisions called compartments within the ectoderm of the embryo. Compartments were originally defined by the property that the cells in a compartment would mingle readily among themselves, but not with cells in adjacent compartments.

Like the alternating nf and ns blast cells in the leech embryo, compartments arise along the length of the fly embryo at exactly twice the spatial frequency of the segments. Each metameric repeat therefore contains two compartments, which can therefore be defined as "anterior" and "posterior" compartments on the basis of their position within the traditionally defined segments.

A variety of experiments indicate that the "parasegments" generated by pairing each "posterior" compartment with the *caudally* located "anterior" compartment are of greater embryological significance. Each compartment is a polyclone and extends across the entire epidermis.

The similarity at the cellular level between the alternating nf and ns blast cells in leech and the alternation of compartments in fly is obvious, but whether this represents homology is less clear-cut. For example, leech segments are not polyclones founded by seven bilateral pairs of primary blast cells, because there is extensive interdigitation of adjacent clones, especially within the M, O, and P lineages. These complications notwithstanding, the observation that two primary blast cells in each N (and Q) cell line are required to make one segment's worth of definitive progeny had led, at least indirectly, to more promising comparisons between leech and fly at both the cellular and the molecular levels.

Early lineage-tracing experiments showed that in normal development, the two n blast cells contributing to a given segment make distinct sets of definitive progeny, mainly neurons of the segmental ganglion. The blast cell clones that arise from alternate n blast cells can also be easily distinguished by their morphology in the germinal plate during stages 8 and 9. Similar observations hold for the q blast cells, which mainly contribute dorsal epidermal progeny to the segment. Thus, the N and Q teloblasts constitute an intriguing variant of the stem cell division mode termed the "grandparental iteration". Various ablation experiments failed to reveal the plasticity in the fates of n blast cells that would be expected if their fates were determined by cell interactions within the germinal band. While such negative evidence can never be convincing, several different sets of observations, summarized here, lend support to the alternative idea that primary blast cells have distinct identities from birth. Stereotyped differences in the timing and symmetry of early mitoses of the primary blast cells provide evidence that these cells are different from one another, at least by the time they enter their first mitoses. For example, the class of primary n blast cells designated nf has a cell cycle of 20 hr duration and then divides unequally to generate larger anterior and smaller posterior secondary blast cells. Alternate primary n blast cells of the class designated ns, have a cell cycle of about 28 hr, divide equally, and contribute Mostly to the anterior portion of the ganglion. Since all the primary ectodermal blast cells lack G1 phase and have S phases of about the same length (about 4 hr), this analysis provided no earlier time point by which the cells were obviously different. For technical reasons, it is impossible to learn the precise composition of the cell cycles by which the parent teloblast generates nf and ns blast cells using the BrdUTP labelling technique.

If N teloblasts in a group of embryos are injected with lineage tracer at random times during stage 7, after they have begun producing primary blast cells, the most anterior-labelled primary blast cell is equally its likely to be of the nf class its of the ns class. This observation indicates that the cell cycles by which nf and ns blast cells are produced are equal in overall length. Moreover, microinjected cell lineage tracers diffuse throughout the injected teloblast in minutes: and because the relatively large volume of the teloblast acts as a reservoir of lineage tracer, the primary blast cells to which the teloblast subsequently gives rise are usually labelled with equal intensity. In a significant proportion of embryos, the first blast cell produced after the injection is labelled with markedly less intensity than all the subsequent cells. For brevity these faintly labelled cells are called faint first blast cells. The exact cause of this phenomenon is unknown: one possibility is that it results when the tracer is injected near the end of cytokinesis, before the membranes have completely sealed, but when the presence of the midbody prevents ready diffusion between the two compartments.

A priori one would express that these faint first blast cells would be equally likely to be of the nf and ns classes. But faint first blast cells are in fact more common when the first labelled blast cell is of the nf class than when it is of the ns class. This result could reflect a longer telophase in cell cycles generating nf blast cells than in cell cycles generating ns blast cells.

The process of regionalization and segmentation in *Drosophila* can be divided into five semidistinct steps as defined by the action of different genetic syntagmata. First, maternal regulatory gene products such as *bicoid* are deposited asymmetrically along the anterior-posterior (A-P) axis in the egg. These products determine where the second set of genes, that define broad regions along the A-P axis, the gap genes, are expressed in the syncytial blastoderm.

Before cellularization, interactions between the gap genes and the maternal regulatory gene products allow the third set of genes, that control segment number, the pair-rule genes, to be expressed in metamerically repeated stripes along the A-P axis, separated by a single segment's width. The members of a fourth set of genes, called the segment polarity genes, operate at various times, each within a specific portion of a given segment, to establish A-P polarity.

After cellularization of the blastoderm, further interactions between the pair-rule and gap gene products determine where along the A-P axis each of the fifth set of genes, the homeotic genes, is expressed. These genes affect regionalization, i.e., individual segment identity, by interacting with target realizator genes.

A frequent characteristic of homeotic genes in *Drosophila* is that the domain of action and expression lies posterior to some rostrocaudal boundary, the exact position of which varies from one homeotic gene to another.

One of the most dramatic, and yet to be understood, aspects of the homeotic genes is that they are physically arrayed along the chromosome within a single complex in the order reflecting the rostrocaudal progression of the boundaries defined by their expression.

Moreover, not only are homologous of these genes found in other phyla, including chordates, but this same correlation between expression pattern and chromosomal arrangement holds true. In light of the foregoing considerations, it is hardly surprising that several candidates for homologous of homeotic genes have been identified in leeches. To date, the expression of only one of these genes, *Lox*-2, has been published. This gene bears equally close sequence homology to two Drosophila homeotic genes, *Ultrabithorax (Ubx)* and *abdominal-A* (*abdA*). In *Drosophila* embryos, transcripts of the homeotic genes begin to accumulate at approximately the same time as do those of *en*, but *Lox*-2 transcripts accumulate in late stage 8 and stage 9, around the time when germinal plate formation is completed and markedly later than the early expression of *ht-en* protein. Nonetheless, the pattern of accumulation shows striking parallels to that of homeotic genes in *Drosophila*. *Lox*-2 is expressed in all five bandlets, except possibly the p bandlet, beginning in segment M6 or M7 and extending throughout all (or most, depending on the species of leech examined) of the more posterior segments.

Given the analogy between the subdivisions of the N (and Q) lineage in leech embryos and the anterior/ posterior compartments in *Drosophila*, one obvious place to look for homology in their segmentation processes was with the *Drosophila* segment polarity gene called *engrailed* (en). At about the same time that compartment boundaries of established in *Drosophila*, it circumferential hand of cells corresponding to the posterior compartment of each segment begins expressing the gene en.

As one would expect when dealing with interacting regulatory genes, the assumption of anterior and posterior compartment fates does not reduce to simply whether the cells express en. But *Drosophila* embryos mutant for *en* do show signs that the identity of the posterior compartment has been perturbed; restrictions on cell mingling across presumptive compartment boundaries are lost and cuticular structures appropriate for the anterior compartment are seen in the presumptive posterior compartment.

Because *en* homologous had been cloned from a number of animal groups, including even vertebrates, it seemed likely that a homologue should be present in leech, and this proved to be the case. The expression of this *engrailed*-class gene (*ht-en*) was studied with the expectation or, more accurately, the hope that it might be a marker for the early differentiation of the two classes of blast cells produced in the n (and/or q) bandlets, e.g., nf and ns primary blast cells. Since it had been shown that primary blast cells initiate transcription immediately upon their birth, it was hoped that *ht-en* would be expressed by alternate blast cells and their descendant clones in the n and q bandlets from the time of their birth, or even by the parent teloblast during every other cell cycle. The expression pattern of *ht-en* was initially misinterpreted through the use of a monoclonal antibody that cross-reacts with engrailed homologous in many, but not all, species. Leeches are among the exceptions. Using a polyclonal antibody raised against a translation product of the leech gene, investigators have described in some detail the true expression pattern of *ht-en*. *Ht-en* is expressed in a highly stereotyped, segmentally iterated pattern, both in early development before gangliogenesis and later during neurogenesis. In *Helobdella*, it is clear that many of these cells expressing *ht-en* later are not clonal descendants of cells that initially express *ht-en* (a point that remains unresolved for cells expressing *engrailed* in later *Drosophila* development). The later expression is predominantly, if not exclusively, in segmentally iterated neurons and appears to persist into adulthood.

Before gangliogenesis, *ht-en* is expressed in a specific subset of cells in each of the bandlets, including the mesodermal (m) bandlet and the two ectodermal bandlets (o and p), in which each primary blast cell makes one segment's worth of progeny.

The timing of *ht-en* expression varies among the five types of bandlets; within each bandlet the timing of expression is invariant with respect to age of the individual blast cell clones and therefore occurs in a rostrocaudal progression that mirrors the birth order of the blast cells, *ht-en* is not expressed by any of the primary blast cells or teloblasts. The earliest expression (e.g. in the p bandlet) is in blast cell clones comprising at least five cells. When cells expressing *ht-en* divide, *ht-en* may be expressed by one, both, or neither of the daughter cells. The early expression of *ht-en* occurs in a transverse stripe of cells that defines the boundary between prospective ganglia. This stripe is most obvious in the cells that express *ht-en* in the N lineage, but all the cells that express *ht-en* during this early period line up within a narrow rostrocaudal domain of the prospective segment. In the n and q bandlets, the early expression of *ht-en* is confined to just one of the two types of blast cells. In the n bandlet, the transverse stripe of cells that express *ht-en* are members of the nf clone. In the q bandlet, the entire qf clone expresses *ht-en* during the period when the clone consists of four or five cells.

By comparing the expression patterns of *engrailed*-class genes in various organisms, we can try to infer the common features that would have been present in their common ancestor. Until techniques are devised for manipulating the expression of *ht-en* in leech embryos, these comparisons can also guide speculation about the function of *ht-en* in leech development. The features of engrailed-class gene expression that appear to be common to leeches and flies include expression both before gangliogenesis and during neurogenesis in unrelated sets of cells.

Early expression is known to play a role in segmentation in fly and is at least correlated with the formation of interganglionic boundaries in leech. Although the definitive segments in the leech body do not constitute polyclones and cannot be divided into compartments as in *Drosophila*, the alignment between bandlets of the cells that express *ht-en* Suggests the possibility of a transient grouping of clones in the germinal plate during stages 8 and 9 that is homologous to the *Drosophila* parasegment. The expression of en defined the posterior compartment throughout the lifetime of *Drosophila* and that the compartment boundaries were maintained by the clonal inheritance of en expression from *en*$^+$ cells in early embryogenesis. Cells expressing *en* can divide and give rise to *en*$^-$ cells, and that in later development en cells are found within the boundaries of the posterior compartment. Therefore, in *Drosophila*, early *en* expression is not clonally transmitted, and *en* expression is not required in the entire compartment throughout the lifetime of the animal. A similar transient and nonclonally transmitted expression of *ht-en* was initially observed in leech. These features may be general features of the expression of *engrailed*-class genes in protostomes since other arthropods (crustaceans) do not show strict clonal transmission of en expression. Scholtz *et al.* found that a subset of *en*-expressing cells in a clone loses en expression, but that a partial subset of the nonexpressing cells later regains expression. The finding that *ht-en* i, expressed in the mesoderm of *Helobdella* contrasts with the findings in *Drosophila.* But mesodermal expression of *engrailed*-class genes is seen in other distantly related groups, including chordates and onychophorans. Thus, it seems likely that *engrailed* was expressed in both mesoderm and ectoderm of the common ancestor of these groups (if the common ancestor had mesoderm) and that it was lost from the mesoderm of arthropods after the annelid and arthropod lines separated in evolution.

The similarity of expression patterns for *engrailed*-class genes in annelids and arthropods, the syntagma concept can be investigated by looking for annelid homologous of genes that interact with *en* in *Drosophila* and characterizing their expression patterns. Two genes that interact with *en* in *Drosophila* are *wingless* (*wg*) and *cubitus interruptus Dominant* (*ci-D*): in the cellular blastoderm of *Drosophila, wg* is expressed in the band of cells just anterior to those expressing en in each segment, and *ci-D* is eventually expressed in all the cells except those expressing *en*. These genes are also segment polarity genes and (with others) can be considered as belonging to the same syntagma as *en*. This syntagma appears to function also in the development of the wing imaginal disc, presumably by operating on a different set of realizator genes. Both of these genes have homologous in other species, including vertebrates.

In leech, a clear sequence homologue to *ci-D* (designated *Htr-ci-*1) has been identified by PCR, and a sequence homologue to *wnt-*1 may have been found. Given the detailed knowledge of the expression patterns of ht-en, clear predictions as to how these genes would be expressed if their syntagmal relationships were conserved can now be made.

Other possible leech homologous of *Drosophila* homeotic genes have also been isolated, and their expression patterns are being analysed; these include *Lox*5, an *Antennapedia*-like gene; *Lox*6, a *Deformed*-like gene; and *Lox*7, a *labial*-like gene. Obvious questions are how many and what other homologous of what is now termed the HOM/Hox complex will be present in leech and whether their organization along the chromosome(s) will also be conserved. Blast cells fates are specified one by one as they arise from the parent teloblast. Thus, understanding how this cell cycle-dependent specification of cell fate proceeds presumably with the involvement of homologous of the same genes that control regionalization in *Drosophila*, becomes the crux of the distinction between the spatial and temporal modes of regionalization and segmentation. For the rostrocaudal patterning gene homologous that have been described to date in leech, expression occurs only after proliferation of secondary blast cells is well advanced and thus gives no definitive information about processes in the teloblasts. As yet, no leech homologous to *Drosophila* genes in the maternal, gap, or pair-rule classes of developmental regulatory genes have been identified. But assuming these classes of genes do exist in leech, can the results already obtained with the homologous of segment polarity and homeotic genes guide our thinking about what to anticipate for their occurrence and function?

One means by which a cell cycle-dependent specification of blast cell fates might be achieved would be for the teloblast to undergo a heritable change with each cell cycle. Thus, each blast cells would receive an inheritance reflecting the state of the teloblast during the cell cycle of its birth. The results obtained so far suggest that the state of the teloblast might be dictated by the qualitative and quantitative mix of factors that regulate segment polarity and homeotic genes. For example, a *bicoid* homologue translated from a maternally inherited mRNA (if such exists) could be present at high levels in the newborn teloblast, and then decline continuously during blast cell production as the *bicoid*-homologue mRNA and protein are either degraded and/or partitioned into blast cell progeny. If homologous of gap genes were also present in leech and functioning in an evolutionarily conserved syntagma with the postulated *bicoid* homologue, we could further predict that a young teloblast would also contain high levels of *hunchback* homologue, whose expression would fall as bicoid levels dropped, permitting the expression of *Kruppel* and *Knirps* homologous to rise and fall sequentially within the teloblast and their descendant blast cells. Each chain of blast cells would be thereby subdivided into transcriptionally distinct groups of cells.

The temporal control of segmentation is relatively straightforward in the case of the O, P, and M lineages. Since individual blast cells produced by M and O/P teloblasts give rise to one segmental complement of progeny, the simple reiteration of (approximately) the same developmental program by each blast cell produces a segmentally repeated pattern of tissues along the anterior-posterior body axis. The behaviour of the N and Q lineages is more complex and appears to require an additional level of regulation, however.

Each n and q bandlet consists of alternating cell types (nf and ns; qf and qs), so that for these bandlets, one segmental complement is descended from a pair of blast cells. As described here, it appears that the f and s blast cells differ at birth; moreover, since *ht-en* (the leech homologue of *engrailed*) is not expressed until the s and f blast cell clones are well along their separate developmental pathways, it seems unlikely that a putative syntagma of segment polarity gene homologous in leech is involved in the initial specification of different fates for f and s blast cells. Another possibility is that "upstream" regulators of the segment polarity genes are involved in distinguishing the s from f cells; if we persist in seeking homology with Drosophila, homologous of the pair-rule genes would be logical candidates for Such upstream regulators.

The large promoters of genes contain many relatively independent regulatory sites, each of which is capable of initiating expression in response to a particular concentration of bicoid and gap gene products. Thus, these promoters "read" the position of the nucleus along the length of the embryo. Other secondary pair-rule genes, such as *fushi tarazu*, are regulated by the primary genes. In leech, the production of f and s class cells could result from a further refinement of the heritable states through which the teloblast progresses, due to cyclic changes in the concentration (or activity) of one or more pair-rule gene products. Any such scheme requires that the pair-rule gene products be relatively short lived and also that the effects of pair-rule gene expression be reversible in some sense, so that the stale of the teloblast can be "reset" from the state in which one pair-rule gene is on to the one in which it is off and vice versa.

The cyclic changes in activity of the pair-rule genes would result from interactions among the products of the pair-rule genes themselves and with other genes, including presumably the gap genes, and one or more of the genes involved in controlling the cell cycle of the teloblast. Interaction with the cell cycle control mechanism is postulated to account for the fact that f and s blast cells arise in strict alternation during successive cell cycles of the N and Q teloblasts.

The postulated network of interacting genes would constitute a biochemical oscillator. Activities of the components of such oscillators (in this case, presumably the various gene products) rise and fall in defined phase relationships. Here, one or more of these cycling gene products would also serve to determine the different states of successive blast cell progeny. In the simplest form of this model, a single pair-rule gene product would be expressed, during alternate cell cycles in the N and Q teloblasts.

Two or more pair-rule genes functioning within the teloblast, each with fewer but longer periods of overlapping expression, which define distinct states of the teloblast in a combinatorial manner. This latter model is appealing because it more closely resembles the current proposals for the role of pair-rule genes in specifying cell fates in the *Drosophila* blastoderm. But evolutionary arguments notwithstanding, we must also bear in mind that this aspect of leech development may be controlled by a genetic mechanism that is unrelated to those involved in *Drosophila* or by a mechanism that is not directly related to gene expression at all.

The development of the germ line in the mouse embryo involves a number of distinct processes or phases leading to the establishment of the meiotic population. Primordial germ cells (PGCs), the embryonic precursors of the sperm and ova of the adult animal, arise as a small population of cells in the early postimplantation embryo at about 7 days postcoitus (dpc).

The PGCs are classically identified as cells expressing alkaline phosphatase and are thought to represent a subpopulation of the total number of PGCs present in the embryo at that time. During the subsequent 5 or 6 days of embryonic development. PGC numbers increase dramatically and they migrate to the gonad anlagen. Once in the gonad, male germ cells enter mitotic arrest while female germ cells enter directly Into meiosis. Thus the establishment of the germ line in the embryo involves the segregation of PGCs from the somatic lineages, proliferation of PGCs, migration of PGCs to the gonad anlagen, and finally differentiation in the gonad. These processes are of interest in their own right but are also used as model systems to study cell proliferation, cell migration, and cell differentiation. Surprisingly, although the survival of the germ line is essential to the survival of animal species, until recently very little was known about the factors regulating these distinct phases of PGC development.

PGCs are classically identified as alkaline phosphatase-positive cells and are first identified as a population of approximately eight cells in the extraembryonic mesoderm of the early 7 dpc embryo. These alkaline phosphatase-positive PGCs probably represent a fraction of the PGCs present in the embryo at that time and which have differentiated from cells originating in the epiblast.

Expression of alkaline phosphatase, therefore, occurs gradually during this period of germ cell differentiation. By 8.0 dpc, approximately 125 alkaline phosphatase-positive PGCs can be identified at the caudal end of the primitive streak near the allantoic rudiment. By this time it is thought that all PGCs express alkaline phosphatase and thus this number represents the total number of PGCs present in the embryo.

During the next 24 hr of development PGCs are incorporated into the hind gut as it invaginates. This movement of PGCs is thought to be a passive process brought about by the morphogenetic movements of the embryo. At this point in their development the PGCs are some distance from the site at which they will undergo their final differentiation into meiotic cells. During the subsequent 4 days of development, PGCs undergo active, directed migration to colonize the gonad anlagen. They migrate out of the hind gut, up the dorsal mesentery of the gut toward two thickened ridges of tissue (the genital ridges or gonad anlage) lying either side of the hind gut mesentery on the dorsal body wall. In the course of migration and following colonization of the gonad anlage, PGC numbers increase dramatically so that by 13.5 dpc they form a population of about 25,000 cells in the gonad. In the male, PGCs undergo mitotic arrest but continue to grow in size. They reenter mitosis about 5 days after birth, forming the mitotic stem cell of the postnatal testis, the spermatogonium. In the female, however, PGC enter directly into meiosis In the embryo, arrest at meiotic prophase, and reenter meiosis a few days after birth. Up to the time of birth considerable loss of female PGCs occurs through a process known as atresia. These differentiation events occur gradually in the PGC population as is the case with the acquisition of alkaline phosphatase activity earlier in their development. Thus not all male PGCs enter mitotic arrest as a cohort nor do all female PGCs enter meiosis simultaneously. Coincident with these differentiation events are changes in PGC surface antigen expression. Many of these cell surface antigens are carbohydrate differentiation antigens and represent post-translational modifications of glycolipids and glycoproteins. Perhaps the best studied of these antigens is the stage-specific embryonic antigen-I (SSEA-1) which represents a fucosylated lactosamine modification of glycoprotein and glycolipid cores. The SSEA-1 antigen is expressed on the PGC surface from 8.5 dpc, through the period of migration, and is lost after PGCs colonize the gonad anlagen. The functional significance of antigenic modulation during the development of PGCs remains unclear but the presence of these antigens on the surface of PGCs represent useful tools for the purification of PGCs from embryonic gonads and their identification in culture. This brief sketch describing the embryonic history of the germ line illustrates some of the questions still outstanding about the emergence of the germ line in animal species: (1) When and where is the germ line set aside in the embryo? (2) How is the germ line segregated from the somatic lineages? (3) What triggers PGCs to begin migration to the gonad anlagen? (4) What factor(s) guides PGCs to the gonad anlagen? (5) Upon gonadal colonization, what factor(s) causes cessation of PGC migration? (6) What factor(s) regulates PGC proliferation? (7) What factor(s) regulates PGC differentiation in the gonad? (8) What factors regulate mitotic or meiotic arrest of male and female PGCs?

In vitro analysis has been used to investigate many of these questions. Moreover, like a number of other migratory embryonic cell types (most notably neural crest cells), PGCs have been used as a model system to study the control of cell migration, and much of this analysis has been carried out using *in vitro* systems. PGCs begin and end migration at a specific time in development and migrate to a single site, the gonad anlagen. Their migration is therefore controlled very strictly both temporally and spatially. Like many tumor cells PGCs migrate through solid tissues rather than through pathways of extracellular matrix that form intercellular spaces. Thus PGCs represent a useful model system by which to study the normal control mechanisms that regulate cell migration in tissues and the breakdown of these control mechanisms during tumor metastasis. The following sections discuss how *in vitro* systems have been used to analyse growth factor regulation of PGC development and what further developments and advances are likely in the near future. Two conditions which alter the normal pathway of PGC development identify genes involved in that process.

A number of mutations in the mouse cause sterility in the homozygous condition, thereby identifying genes involved in germ cell development. Some of these have been shown to act during the period of PGC development, including *Dominant While Spotting (W), Steel (SI). Hertwigs' macrocytic anemia ((in),* and *germ cell deficient (gcd).* The genes encoded at the W and Sl loci have been characterized at the molecular level and will be described in detail. A second situation in which PGC development is altered is when PGCs give rise to tumors, benign teratomas, and malignant teratocarcinomas. In this case the PGC gives rise to a pluripotent stem cell, the embryonal carcinoma (EC) cell, which can differentiate into many cell types representative of all the three primary germ layers.

In teratomas, all the EC cells differentiate, thereby forming a benign tumor whereas teratocarcinomas, some of the EC cells fail to differentiate and continue to proliferate thereby forming a malignant tumor. Some of the genes affecting these processes have been identified by analysing teratoma and teratocarcinoma incidence in inbred strains of mice and congenic derivatives carrying specific gene mutations. The genes identified that affect teratocarcinogenesis include *Sl,* situs inversus viscerum (iv), agouti (A), and teratocarcinoma(ter). Understanding the role of these genes in teratocarcinogenesis may provide useful information as to their role in normal PGC development.

Primordial germ cell culture

PGCs can be cultured in a variety of conditions, including organ culture, feeder cell-dependent culture, and feeder cell-independent culture. Each technique has its uses and problems. The earliest attempts to culture PGCs were aimed at understanding PGC migration. Blandau and colleagues derived squash preparations of fetal gonads in order to study the migratory properties of mouse PGCs. These cultures were not aimed at maintaining PGCs in culture for long periods of time. Although PGCs can be cultured on glass, plastic, agar, or gelatin-coated substrata, their survival is poor.

Moreover, the number of cells that can be recovered from the embryo onto these substrates makes these experimental conditions unfavourable for most purposes. Extracellular matrix components such as fibronectin, laminin, and type IV collagen can improve recovery of PGCs, but their viability remains low.

Most long-term Cultures of PGCs have therefore involved growing them in conjunction with feeder layers of mitotically inactive somatic cells. Heath and Evans both reported that mouse or rat PGCs Could be maintained in culture for about 5 days when cultured on top of feeder layers of their own embryonic somatic cells or fibroblast feeder layers. Nevertheless, even in these conditions, PGC numbers declined over the 5-day culture period, and PGCs did not divide or incorporate tritiated thymidine into their DNA. Although it was not possible at that time to derive long-term cultures of PGCs (or to establish PGC-derived cell lines), various improvements in the culture conditions led to the establishment of cultures in which PGCs could both survive and Proliferate for about 7 days. The use of STO cells or TM4 cells allows culture of PGCs in conditions in which they both survive and proliferate for up to 7 days. PGCs can be identified in feeder layer culture by alkaline phosphatase histochemistry and by reactivity with antibodies directed against cell surface differentiation antigens such as SSEA-1. The major problem with these cultures is their complexity since they contain not only PGCs and feeder cells but also mitotically active embryonic somatic cells. Thus the interpretation of the action of growth factors in these conditions is difficult. Any conclusions about the direct action of it growth factor on PGCs must be supported by a rigorous demonstration that PGCs express a particular growth factor-receptor. The use of feeder layer culture for PGCs gradually led to improvements in PGC culture and the identification of conditions for stimulating PGC proliferation. When isolated into culture, PGCs are initially identified as single cells growing on top of the feeder layer.

After a few days of culture, small groups of PGCs form either by clonal division or by aggregation, These groups enlarge over the next 3 or 4 days of culture but they eventually disappear. Whether they differentiate into an alkaline phosphatase-negative cell population or simply die is unknown. When isolated from the embryo and placed into culture their mitotic period mirrors that seen *in vivo* as does expression of the carbohydrate differentiation antigen SSEA-1. These two observations Suggest that PGCs may differentiate *in vitro* according to their normal schedule *in vivo*. The appearance (rather than the disappearance) of differentiation markers would definitively answer this question but no such markers exist at the present time. Using feeder-dependent culture systems it is possible to investigate factors affecting PGC survival, proliferation, and differentiation.

One important observation from *in vitro* assays was that the serum concentration *could* influence the rate of PGC proliferation-proliferation was greater in medium containing 15% fetal calf serum (FCS) than in 10% FCS.

Another important observation was that cAMP agonists such as forskolin and cholera toxin could greatly. Although these improvements are useful in maintaining PGCs in culture, little was known about the growth factor requirements for PGC survival, proliferation, and differentiation.

Many mutations at both the *W* and *Sl* loci, in addition to their effects on the hemopoietic system and melanoblasts, cause sterility in the mouse, thereby identifying genes involved in germ cell development. *W* and *Sl* mutations are semidominant and can be identified as heterozygotes by their effect on coat pigmentation. A large number of *W* or *Sl* alleles, when homozygous, cause lethality: *W/W* animals die around the time of birth, while *Sl/Sl* animals begin to die at around day 15 of gestation. Most probably this is because of the dramatic effect of these mutations on the hemopoietic system-both *W/W* and *Sl/Sl* animals are severely anemic. In *W* and *Sl* homozygote mutant embryos, PGCs are formed, but fail to divide, and their numbers remain low. The few PGCs that reach the gonad do not survive and the animals are generally sterile.

The *W* locus is now known to encode *c-kit*, a receptor tyrosine kinase (RTK) with homology to the receptors for colony-stimulating factor-1 (CSF-1) and platelet-derived growth factor (PDGF). These receptors are characterized, in part, by having a unique insert in the tyrosine kinase domain. In the embryo, *c-kit* is expressed in cells affected by *W* mutations, that is, hemopoietic stem cells, mast cells, melanoblasts, and PGCs and thus coincides with the cells affected by *W* mutations. A large number of mutations at the *W* locus have been identified, each differing in the severity of their effects oil hemopoietic stem cells, germ cells, and melanoblast. Thus some W alleles are viable and fertile but lack coat melanocytes, while others impair the hemopoietic system, are severely anemic, and are perinatal lethals.

The molecular bases of many W mutations have been characterized. Severe *W* mutations (i.e., those causing lethality), such its W^{42} are characterized by having mutations or deletions in the *c-kit* kinase domain and have impaired or no *c-kit* kinase activity. Less severe *W* alleles (viable), such as *W-sash* W^{sh}), are regulatory mutants that affect the level rather than the activity of *c-kit* protein.

Evaluation of many weak, moderate, and severe *W* mutations reveals a correlation between the degree of severity in homozygotes and the level of c-kit kinase activity. Thus the simplest explanation for the sterility seen in W mutant mice is an impaired c-kit signalling pathway in PGCs. The *Sl* locus is now known to encode the ligand for *c-kit,* known as mast cell growth factor (MGF), kit-ligand (KL), stem cell factor (SCF), and Steel Factor (SLF). For the purposes of this chapter this factor is referred to as Steel Factor. SLF is a transmembrane growth factor that exists in two forms generated by alternate splicing. One of these forms can be efficiently cleaved to release the extracellular domain as a soluble factor while the other form is presumed to remain membrane bound. *In situ* hybridization analysis reveals that SLF mRNA is expressed not in the cells affected by *Sl* mutations (hemopoietic stem cells, germ cells, and melanoblasts) themselves, but rather in the surrounding tissues. This is entirely consistent with earlier transplantation and grafting studies demonstrating that the Sl gene product acts in the cellular environment.

A number of *Sl* mutations have now been characterized. Generally, the most severe *Sl* mutations (lethal), such as *Sl*, Sl^{l}, Sl^{12H}, *Sl*-grizzle belly and Sl^{18H}, represent complete deletions of the *Sl* gene, while viable

Sl alleles, such as Sl^{d}, Sl^{17H}, and *Sl-Panda* (Sl^{Pan}), involve small deletions or alterations in the *Sl* coding sequence or in genomic regulatory regions. What role does SLF play in PGC development? In principle SLF could have a role in PGC survival, proliferation, migration, or differentiation. Four pieces of evidence suggest that SLF is required for PGC survival. First, the role of SLF expression by somatic cells was examined in *in vitro* assays. In culture, PGCs will survive for up to 7 days when grown on feeder layers of certain fibroblasts.

The ability of PGCs to survive and proliferate on fibroblasts is dependent in part, on the production of SLF by these cells. PGCs could not survive for even 24 hr on feeder layers of CV-1 cells that did not produce SLF, although they could adhere to these feeder layers in short-term adhesion assays. When CV-1 cells were reconstituted with SLF by transfection, these reconstituted cells could now support PGCs for up to 3 days. Thus SLF is both necessary and sufficient for PGC survival in culture. Second, when a soluble, recombinant form of SLF (rSLF) was added to PGCs grown on STO cells, it increased both the numbers of PGCs and size of PGC colonies in culture. However, in these conditions rSLF did not stimulate PGC proliferation as assessed by bromodeoxyuridine (BrdU) labelling. Similarly, addition of rSLF was unable to stimulate PGC proliferation on NIH-3T3 feeder layers. When rSLF was added to PGCs grown in the absence of feeder cells on gelatin-coated substrata it increased PGC survival and was unable to stimulate proliferation. Third, signalling via the *c-kit* receptor was blocked by the addition of an *anti-c-kit* antibody (ACK2). This antibody not only blocks PGC proliferation in culture but also seems to inhibit PGC survival. These data can be interpreted to show that SLF is not a mitogen by itself but is a PGC survival factor. Fourth, rSLF suppresses apoptosis (programmed cell death) of PGCs in culture consistent with the notion that the c-kit signalling pathway is required for PGC survival. This is a some what controversial issue since rSLF can stimulate PGC proliferation on feeder cells that themselves do not express SLF, although PGC survival in these conditions is short lived. Distinguishing between proliferation and survival in feeder-dependent cultures may be very difficult.

An important point to bear in mind, however, is that addition of a survival factor (that itself is not a mitogen) may allow cells to respond more efficiently to available mitogens. Thus addition of rSLF may improve PGC survival sufficiently to allow the cells to respond to other factors. Moreover, factors which can promote cell survival may, in the presence of other factors, act as mitogens.

For example, neurotrophin-3 (NT-3) alone can promote short-term survival of oligodendrocytes and their precursors, but in combination with PDGF and ciliary neurotrophic factor (CNTF) (or related factors) it can promote long-term survival and also stimulate proliferation. Evidence for SLF acting in other ways, for example, as a chemoattractant, remains inconclusive. In *in vitro* assays rSLF was unable to act as a chemoattractant but these assays were carried out in the presence of STO cells which produce SLF, so interpretation of these data is complicated. Evidence for SLF acting as a PGC adhesion molecule will be presented next. In summary, SLF seems to be required for long-term survival of PGCs in culture and together with other factors, can stimulate PGC proliferation.

SLF exists in two forms generated by alternate splicing. One form is predicted to be a membrane-bound factor whereas the other form can be cleaved to generate a soluble factor. Important questions include why two different forms of SLF exist and whether there are differences in their effect on PGCs? Evidence for the importance of membrane-bound SLF in PGC development comes not only from cell culture analyses but also from the analysis of two *SI* mutations, *Sl* and Sl^{17H}. The Sl^d allele has been characterized at the molecular level and results from a discrete 4.0-kb intragenic deletion that removes sequences encoding the SLF cytoplasmic tail and transmembrane region. Thus Sl^d/Sl^d (or Sl/Sl^d) animals can produce only a soluble form of SLF.

The sterility seen in Sl/Sl^d animals suggests that soluble SLF encoded by the Sl^d allele may not be sufficient to support PGC survival. This hypothesis was tested in *in vitro* culture by transfecting CV-1 cells with a gene encoding the Sl^d allele. These cells could support limited PGC survival for only 24 hr whereas cells transfected with the wild-type allele could support PGC survival for at least 3 days. Similar experiments were carried out using two bone marrow-derived cell lines generated from congenic lines of mice. One cell line (BM-WT) produces membrane-bound factor, while the other cell line (Sl/Sl^d) produces only soluble factor.

Once again, PGCs could only survive on the cells expressing the full-length, membrane-bound factor. Addition of further soluble rSLF improved initial survival over 24 hr but was unable to effect long-term survival of PGCs. Similarly, Hogan and colleagues have demonstrated that membrane-bound factor is more efficient at supporting PGC proliferation. PGC proliferation was assayed in the presence of saturating amounts of soluble, recombinant SLF on a SLF-null cell line or on a recombinant derivative cell line (Sl-m220) expressing a noncleavable from of SLF. The conclusion drawn from this experiment was that even in the presence of saturating amounts of Soluble SLF, PGC proliferation was greater in the presence of membrane-hound SLF arid therefore that membrane-bound SLF is required in addition to soluble SLF for optimal effects on PGCs. Studies on the effect of SLF on hemopoietic stem cells have also shown that transmembrane forms of the factor enhance stem cell survival and proliferation.

Since *Sl*^{d}/*Sl*^{d} animals lack all coat melanocytes, these data suggest that, like PGCs, melanocytes require full-length transmembrane forms of SLF for their survival and/or proliferation *in vivo.* In culture rSLF is required for the maintenance but not the differentiation of melanoblasts, while analysis of melanoblast development in *Sl*^{d}/*Sl*^{d} embryos shows that transmembrane forms of SLF are required for melanoblast survival. Further evidence for a role for the cytoplasmic tail of SLF comes from the analysis of another mutant *Sl* allele, *Sl*^{17H}. This ethyl nitrosourea (ENU)-induced *Sl* allele is of interest because of its differential effect on the fertility of males and females. Heterozygotes have a lightening of the coat with a pronounced headspot but otherwise are viable and fertile. Homozygotes, on the other hand, are black-eyed but have a white coat. They are fully viable but are fertile only in the female. Molecular analysis of the *Sl*^{17H} allele revealed the complete absence of exon 8 from *Sl*^{17H} mRNA. Exon 8 encodes 23 of the 36 amino acids of the SLF cytoplasmic tail and begins 1 amino acid C-terminal to the transmembrane domain. Thus the SLF encoded by the *Sl*^{17H} allele encodes a protein with identity to the wildtype protein from the N terminus up to the first amino acid of the cytoplasmic tail.

The rest of the *Sl*^{17H} SLF cytoplasmic tail is produced from an alternate reading frame that encodes 27 amino acids before reaching a stop codon. The cytoplasmic tail of the SLF encoded by the *Sl*^{17H} allele is, therefore, not only substantially different from the wild-type protein but also shorter. Two explanations seem possible for this mutation: a point mutation in the 3' splice acceptor site of intron 7 or a deletion of exon 8. Analysis of *Sl*^{17H} DNA revealed a T → A transversion in the splice acceptor site of intron 7 by comparison with wild-type DNA and DNA from other strains of mice, confirming the former possibility. Although the protein encoded by the *Sl*^{17H} allele is altered in its cytoplasmic tail, it is biologically active and can stimulate mast cell proliferation. In terms of its activity on PGCs it seems to fall somewhere between that of wild-type protein and the protein encoded by the *Sl*^{d} allele or soluble, recombinant protein. How then can the male sterility of *Sl*^{17H} animals be explained?

Surprisingly, when *Sl*^{17H} homozygote embryos were examined, it was found that PGC numbers were equally reduced in both males and females even though adult females are fertile. Examination of postnatal and adult gonads of *Sl*^{17H} homozygotes identified a defect in spermatogenesis but seemingly no effect on oogenesis apart from a reduction in the number of oocytes. This reduction is most likely to be due to the reduction in numbers of PGCs entering and surviving in the embryonic gonad. Over time all the oocytes and follicular tissue seem to be lost from the ovaries of adult *Sl*^{17H}/*Sl*^{17H} animals and the ovary shows invaginations of the surface epithelium and tubular adenoma. This tumor phenotype is also observed in germ cell-deficient *W*^{x}/*W*^{v} females and is thought to be due to overproduction of pituitary gonadotrophins. Following colonization of the gonad, germ cell development in male mice is quite distinct from germ cell development in females. While female PGCs enter directly into meiosis in the embryo, male PGCs enter mitotic arrest.

In the postnatal animal these mitotically arrested cells, now called *gonia*, begin to divide mitotically again about 5 days after birth and establish a stem cell population. These spermatogonia give rise to spermatocytes that divide meiotically generating haploid cells.

The final stage of male germ cell development, spermiogenesis, involves the complex morphogenetic changes that result in the formation of the spermatozoon. How does the Sl^{17H} mutation affect this process? In wild-type animals at 12 days of age all testis tubules contain spermatogoma and 90% of tubules show all stages of germ cells up to meiotic prophase. In Sl^{17H} homozygotes, on the other hand, 12-day-old postnatal testes lacked spermatogonia in 50% of their testis tubules. While some tubules (27%) contained spermatogonia, in only 23% of tubules had germ cells advanced to the spermatocyte stage. Germ cells in the testes of Sl^{17H} homozygote males continue to develop, such that by 5 weeks of age 52% of the tubules contain germ cells advanced to the elongating or maturing spermatocytc. However, these tubules are severely deficient in spermatocytes, indicating that the next round of spermatogenesis has failed to occur. In 8-week-old animals most of the tubules in mutant homozygotes were completely devoid of germ cells with only a few spermatogonia remaining. Spermatogenesis in these animals is therefore late in inception and early in cessation.

That membrane-bound forms of SLF are more efficient at supporting PGC survival and/or proliferation is one explanation for the sterility found in Sl/Sl^{d} animals. Thus it seems likely that the sterility seen in Sl^{d} mice is due to a specific role for transmembrane forms of SLF. The inability of the soluble form of SLF to support long-term survival of PGCs *in vivo* or *in vitro* could imply (1) a role for membrane-bound SLF in a prolonged receptor-ligand interaction, (2) a requirement for it localized high concentration of SLF, (3) a particular conformation of SLF precluded by soluble factor, or (4) a role for SLF in promoting cell adhesion. A requirement for a transmembrane factor could provide a mechanism for PGC guidance to the gonad anlagen along a haptotactic gradient of SLF. SLF mRNA is expressed in somatic cells along the germ cell migratory route and, interestingly, Keshet and co-workers have suggested that a gradient of SLF mRNA may exist along the pathway. However, without knowing precisely where the SLF protein is expressed and localized or how its cleavage is regulated, these data must be interpreted with caution.

One piece of data that is pertinent to this point is the analysis of PGC migration in Sl/Sl^d animals. Since the *Sl* allele is completely deleted for SLF coding sequences, Sl/Sl^d animals can only produce soluble SLF. As adults these animals are sterile, but PGCs are produced in Sl/Sl^d embryos and migrate to the genital ridge. Analysis of PGC migration in these embryos reveals that many PGCs are not on the correct migratory pathway at 11.5 days of development. In wild-type embryos at this stage approximately 70% of PGCs have reached the genital ridge and the remaining 30% are at the coelomic angle (the junction between the mesentery and the dorsal body wall). In the embryos identified as Sl/Sl^d compound heterozygotes, only 24% of PGCs have reached the gonad while 40% are ectopic to the normal pathway of migration. While the authors did not present data describing the positions of aberrantly migrating PGCs, it will be interesting to analyse PGC migration in Sl^d animals now that the molecular basis of this locus is understood. Analysis of PGC migration in other *Sl* alleles will also be of interest.

Similar studies of PGC proliferation and migration in W^e/W^e embryos also demonstrate that many PGCs are found ectopic to the normal pathway of migration. Moreover, these studies demonstrated that many PGCs in W^e/W^e embryos were clumped together rather than spread out along the migratory route. These data lend support to the idea that the *c-kit*/SLF interaction could play a role in cell adhesion. In W^e/W^e embryos in which this interaction is altered, adhesion between PGCs could predominate over cell-substratum interactions. The drastic effect of the Sl^{17H} mutation suggests an important role for the SLF cytoplasmic tail not only in PGC development but also in postnatal male germ cell development.

The SLF cytoplasmic tail could be important for a number of different functions. Dimerization is required for activation of the *c-kit* RTK, and it is possible that SLF also needs to dimerize to function correctly and that the cytoplasmic tail plays a key role in this process. It is also possible that the SLF tail is involved in signalling inside the cells expressing it. Therefore, when SLF and *c-kit* interact, PGCs would receive a signal via c-kit and somatic cells would receive a signal via SLF. Another possibility is that the amino acids comprising the cytoplasmic tail of SLF are important for stabilizing the protein in the cell membrane. In this regard it is important to note the loss of a number of juxtamembrane lysine residues from the SLF encoded by the Sl^{17H} allele. These charged residues could be responsible for holding the factor in the membrane.

A final possibility is raised by work analysing the function of the cytoplasmic tail of pro-TGFα. Like SLF, pro-TGFα is a transmembrane growth factor that has an extracellular cleavage site that allows proteolysis and release of the extracellular domain. Studies of the structure of pro-TGFα identified a terminal valine residue to be important for ectodomain cleavage of pro-TGFa. Mutation of the terminal valine resulted in a pro-TGFα molecule that could not be efficiently cleaved to release a soluble factor. Domain swapping experiments between pro-TGFα and SLF demonstrated that the same regulatory elements exist in the cytoplasmic tail of SLF. Like pro-TGFα, the C-terminal amino acid of wild-type SLF is a valine residue, and this is altered in the factor encoded by the *Sl*^{17H} allele. The factor encoded by the *Sl*^{17H} allele would therefore be expected to show altered regulation of ectodomain cleavage and possibly is not cleaved at all. Thus developmentally regulated cleavage of SLF may play an important role in PGC and postnatal germ cell development. The differential effects of soluble versus transmembrane SLF *in vitro*, the sterility of *Sl*/*Sl*^{d} animals, and the severe effects of the *Sl*^{17H} mutation on PGC development suggest an important role for transmembrane forms of SLF.

Further studies using antibodies to SLF or mutational analysis of SLF should reveal important new data about the function of soluble and transmembrane forms of this factor. Although the precise action (proliferation versus survival) of SLF on PGCs remains unclear, the analysis of SLF function in *in vitro* systems showed that another factor(s) must be acting on PGCs. One such factor is leukemia inhibitory factor (LIF).

Examination of PGC proliferation on different embryonic fibroblast cell lines demonstrated that PGC proliferation does not correlate well with the amount of SLF produced by the feeder layer. NIH-3T3 cells produce more SLF mRNA and stimulate greater mast cell proliferation than STO cells. Moreover, addition of recombinant SLF to PGCs cultured on NIH-3T3 cells has no effect on PGC proliferation. The conclusion drawn from these data was either that another factor produced by STO cells could stimulate PGC proliferation or alternatively a factor produced by NIH-3T3 cells could inhibit *PGC* proliferation. Addition of STO cell-conditioned medium (STO-CM) to PGCs cultured on NIH-3T3 cells stimulated PGC proliferation, thus making the former possibility more likely. Similarly, STO-CM stimulates proliferation of PGCs cultured on *Sl*-m220 cells. The conclusion drawn from these studies is that STO cells produce, in addition to SLF, another soluble factor that is a PGC mitogen. One of the factors made by STO cells is LIF and indeed LIF stimulates PGC proliferation on *Sl*-m220 cells. It seems likely, therefore, that one soluble, mitogenic activity produced by STO cells is LIF.

LIF is a pleiotropic cytokine produced by a variety of different cells that can act in a variety of different ways on a plethora of other cells. The LIF cDNA encodes a protein with a predicted molecular mass of 20.000 kDa but the purified protein is a disulfide-linked monomer with a molecular mass of 5058 kDa. Known by a variety of different names LIF can stimulate proliferation, induce or inhibit differentiation, and act its a positive or negative cellular activator. How does LIF act on PGCs? On STO cells and *Sl*-m220 cells LIF clearly stimulates PGC proliferation measured by BrdU incorporation. On TM4 cells, on the other hand, LIF promotes PGC survival. Moreover, like SLF, LIF can suppress apoptosis of PGCs in culture.

Previous studies, analysing alkaline phosphatase activity and anti-SSEA-1 reactivity of cultured PGCs, led to the suggestion that PGCs may develop on schedule in culture. If LIF were able to inhibit this differentiation step *in vitro* then the observed effect would, presumably, be increased proliferation since PGCs would not progress to a nonmitotic phase. This action of LIF would not be unprecedented since one of its earliest identified activities was as an inhibitor of ES cell differentiation. The analysis of the effects of LIF on the expression of PGC stage-specific antigens may allow this question to be addressed. Whatever the true function of LIF it clearly acts on PGCs and synergizes with SLF to stimulate PGC proliferation *in vitro*. Because of the complex nature of the culture system it is difficult to determine if LIF acts directly on the PGCs or indirectly via STO cells or embryonic somatic cells. A low-affinity, LIF-binding, LIF receptor (LIFR) was cloned and was found to have homology with gp130, the IL-6 signal tranducer. This receptor, together with gp130, forms a high-affinity LIF receptor capable of transducing signals in response to LIF. PGCs isolated from the genital ridge and labelled with an anti-SSEA-1 antibody can also be labelled with an anti-LIFR antiserum, suggesting that PGCs express a LIF receptor. These data strongly suggest that PGCs are a direct target of LIF action in culture.

Does LIF play any role in PGC development *in vivo*? In the mouse embryo the greatest expression of LIF transcripts occurs at the time of implantation in the endometrial glands of the mother. LIF mRNA expression declines rapidly after implantation. Consistent with these data, female mice lacking LIF (created by targeting the LIF gene by homologous recombination in embryonic stem cells) cannot support blastocyst implantation. When LIF-/LIF-embryos are transplanted into a pseudo-pregnant, wildtype female they develop normally and are viable and fertile. This suggests that the major role of LIF is to support embryo implantation.

Although LIF-deficient mice are fertile in both sexes, it is possible that PGC numbers could be reduced in these animals since mice with severe deficiencies in PGC numbers can still be fully fertile. It will be important, therefore, to determine if PGC numbers are normal in LIF-deficient animals. If PGC numbers are normal in these animals then this suggests that either LIF is not the natural ligand for PGCs or another closely related cytokine can substitute for LIF in the absence of LIF.

LIF is a member of a family of cytokines that are related by primary amino acid sequence, secondary structure, and genomic organization. The related cytokines include interleukin-6 (IL-6), ciliary neurotrophic factor (CNTF), oncostatin M (OSM), and granulocyte-colony-stimulating factor (G-CSF).

Understanding of the pleiotropic functions of LIF was advanced by the molecular characterization of the cell surface receptor for LIF. The LIF receptor comprises a low-affinity, LIF- binding subunit (referred to here as LIFR) and a second signal transducer molecule, gp130.

Activation of the LIF receptor complex requires heterodimerization of LIFR and gp130 and is accompanied by tyrosine phosphorylation of both subunits. As members of an emerging cytokine receptor family, LIFR and gp130 serve as components of bi-or tripartite receptors for multiple cytokines. For example, the LIFR/gp130 complex also functions as a high-affinity receptor for the cytokine OSM. Addition of a third component, the low-affinity receptor for CNTF(CNTFRα), converts the bipartite LIFR/gP130 complex into a high-affinity receptor for CNTF. Thus both LIFR and gpl30 are required for the signal transduction of multiple cytokines, including LIF, OSM, and CNTF. These data provide some explanation for the overlapping functions of these cytokines as well as suggesting that some of these factors may be functionally redundant. It seems likely, therefore, that in the absence of LIF another cytokine could substitute for its proposed function in PGC development. Of the LIF-related cytokines, OSM is the most closely related to LIF and it is localized next to LIF on human chromosome 22.

It *is* predicted that OSM could perform all the functions of LIF as well as some others of its in PGC own. Whether OSM could substitute for LIF in PGC development remains uncertain since the OSM gene has only been identified in humans and simians—a murine homologue has never been cloned. Mice lacking CNTF or IL-6 have been created by gene targeting in ES cells and like LIF-deficient mice these animals are fertile. However, in CNTF- or IL-6-deficient animals other members of this cytokine family could potentially Substitute for the missing cytokine. Animals deficient in two or more cytokines in this family may allow the roles of these factors in PGC development to be determined. At the present time, therefore, the precise role of LIF in PGC development *in vivo* remains unclear.

Although SLF and LIF can synergize to stimulate PGC proliferation in vitro the observed proliferation rate compares poorly with that observed *in vivo*. These observations led to the identification of another factor capable of stimulating PGC proliferation, namely basic fibroblast growth factor (bFGF).

While STO cells can produce LIF, the addition of further recombinant LIF to PGCs cultured on STOs stimulates their proliferation. Even in these conditions PGC proliferation is limited, suggesting that other factors might regulate their growth. Of the numerous factors tested *in vitro*, bFGF showed the greatest stimulation of PGC proliferation both on STO cells and on *Sl-m200* cells. bFGF might stimulate PGC proliferation directly or could act indirectly via stimulation of the feeder layer.

At the time of writing no clear picture has emerged. bFGF can stimulate production of LIF and SI-F by feeder cells but could equally act directly on PGCs via a FGF-receptor (FGFR). Until it *is* known whether PGCs express a FGFR, this question will remain unanswered. Even though bFGF is mitogenic for PGCs, it does not on its own affect the long-term survival of PGCs in culture.

The normal pattern of PGC proliferation in culture involves proliferation as single cells; formation of small clones (5-10 cells), either by clonal division or aggregation, and, finally, disappearance of clones between 7 and 10 days after isolation from the embryo. Binary combination of factors (e.g., SLF and LIF or SLF and bFGF) had no effect on this time course of events. However, when PGCs were cultured in the presence of bFGF, LIF, and SLF, long-term proliferation of PGCs was achieved. On fibroblast feeder layers, and in the presence of these three factors, PGCs form clones (5-10 cells) which initially expand as a cellular monolayer and eventually form multilayered colonies that continue to grow and proliferate. These multilayered colonies can be isolated, trypsinized, and replated to generate clonal cell lines that continue to proliferate in culture. In the presence of bFGF, therefore, PGC proliferation is maintained indefinitely. Because of the similarity between these EG cells and EC and ES cells they have been tested for their pluripotentiality in a number of assays. These assays include teratoma formation in histocompatible adult hosts, embryoid body formation in vitro, and, finally, chimera formation when introduced into genetically marked host blastocysts.

In these assays both EC and ES cell lines are pluripotent, that is they differentiate into tissues derived from all three primary germ layers. The major difference between EC cells and ES cells is the efficiency with which they give rise to germ line chimeras when introduced into the blastocyst.

ES cells show high levels of contribution to the germ line in chimeras while EC cells rarely do so. EG cells (or PGC-derived ES cells) generate teratomas in histocompatible hosts and make embryoid bodies in culture, and P.J. Donovan, unpublished observations). When introduced into genetically marked blastocysts. EG cells contribute to all the tissues of the embryo and, in a growing number of cases, to the germ line. These data are in dramatic contrast to the results obtained when

PGCs are tested in these assays. Although purified populations of PGCs have not been tested experimentally for their ability to give rise to teratomas or teratocarcinomas, clearly they can do so under certain circumstances since spontaneous teratomas are naturally derived from PGCs.

Adding bFGF seems, therefore, to alter not only the morphology and proliferative capacity of PGCs, but also broadens their developmental potential. These observations suggest that under certain circumstances PGCs can be made to alter their normal differentiation pathway and regain developmental pluripotency. This observation may have important consequences for the understanding of the mechanism of teratocarcingenesis in mice and humans. Since SLF, LIF, and bFGF modify the *in vitro* and *in vivo* developmental and proliferative potential of PGCs, a key question is whether PGCs are exposed to LIF or FGF2 (or other fibroblast growth factors) during the course of normal development. Presumably if PGCs were exposed to all three factors *in* vivo they would continue to proliferate and could potentially give rise to teratomas and teratocarcinomas.

In order to clearly understand the potential role of FGFs the most important information will be gained from examining the expression of FGFs or their receptors in somatic and germ cells during development. Since EG cell lines have also been derived from both migratory and gonadal PGCs, this suggests that PGCs of many developmental stages could express a FGF receptor. A second possibility is that the stimulatory effects of these three factors (SLF. LIF, and bFGF) could be modified in vivo by the action of negative regulators such as members of the transforming growth factor (TGF) superfamily. When genital ridges are cultured *in vitro* they produce a factor that has a chemotrophic effect on PGCs and can attract PGCs toward the genital ridge in chemotaxis assays. A similar effect is seen when TGFβ1 is used in place of genital ridge-conditioned medium, suggesting that the chemotrophic factor is TGFβ1.

Moreover, antibodies to TGFβ1 inhibit the chemotrophic effect of genital ridge-conditioned medium. Analysis of TGFβ1 expression in the embryo by indirect immunocyto chemistry also reveals the presence of TGFβ1 in the genital ridge. In addition to its chemotrophic properties, TGFβ1 is also it potent inhibitor of PGC proliferation.

Again it is important to note that interpretation of the role of TGFβ1 in culture is complicated by the presence of feeder cells and embryonic somatic cells. Direct action of TGFβ1 on PGCs could be demonstrated if it truly feeder-independent culture of PGCs becomes possible or by demonstration that PGCs express a TGFβ1 receptor. Nevertheless, the mitogenic effects of SLF, LIF, and bFGF on PGCs could be constrained at the appropriate time in development through negative regulation (directly or indirectly) by TGFβ1. Mice lacking TGFβ1 will be an important tool in elucidating the role of this growth factor in PGC development. Another candidate negative regulator of PGC development is another member of the TGFβ superfamily, namely Mullerian inhibitory substance (MIS).

Granulosa cells in the ovary also express MIS but expression in this case must be regulated by another mechanism. Female mice overexpressing MIS are sterile, suggesting that MIS might act directly on germ cells, but its direct action on PGCs in culture has not been determined. Whether or not TGFβ or MIS act as negative regulators of PGC growth in the embryo, it seems probable that negative regulation of PGC proliferation plays an important part in PGC development in mice.

How does bFGF act to modify the proliferative and developmental potential of PGCs? One possibility is that bFGF is simply a PGC mitogen and that LIF (or another factor present in the culture system) modifies PGC growth in culture by affecting PGC differentiation. The analysis of the role of bFGF in oligodendrocyte precursor proliferation suggests another possibility, namely that bFGF blocks differentiation. Oligodendrocyte Type-2 astrocyte (O-2A) progenitors proliferate in response to PDGF, but after a set number of divisions they differentiate on schedule in vitro and lose their ability to respond to PDGF.

It has been proposed that PDGF drives the clock that times oligodendrocyte development in culture. In the presence of bFGF, O-2A progenitors continue to divide and do not differentiate. One effect of bFGF is to increase the steady-state level of PDGF receptors on O-2A progenitors and to increase the sensitivity of O-2A progenitors to PDGF. It has been suggested that bFGF (or a related factor) could induce the PDGF receptor in O-2A progenitors in vivo and set the clock that times oligodendrocyte development. In culture, therefore, bFGF may block O-2A progenitor differentiation by maintaining responsiveness to PDGF.

The parallels between O-2A progenitor development and PGC development are striking. It is intriguing to speculate that exposure of early PGCs (SLF-independent) to FGF4 in the primitive streak might upregulate *c-kit* levels in PGCs initiating a *c-kit*-dependent clock that could regulate the timing of their differentiation. *In vivo* c-kit transcripts are downregulated in PGCs once they have reached the gonad anlagen, and in culture PGCs may follow their normal schedule of development. In culture this could involve PGCs downregulating *c-kit* transcript, and differentiating into a nonproliferating cell. bFGF could inhibit this process by maintaining high levels of *c-kit*, in PGCs and thus lead to long-term proliferation of PGCs. The fact that EG cells, once established, no longer require addition of bFGF and can be grown in the absence of SLF suggests that autocrine mechanisms of receptor activation may become established in these cells. Whatever the mechanism of bFGF action, derivation of EG cells directly from PGCs has a number of important consequences. LIF is still the only factor known to be required for ES cell derivation. Because of this fact it may be possible to extend this technology to other animal species, particularly other mammals.

The generation of bovine, porcine, ovine, or equine EG cells could have great commercial potential and value. The generation of human EG lines, while having major ethical consideration, could be of immense medical benefit in, for example, tissue transplantation and repair. Since ES cell technology is so well established in the mouse, EG cell lines may prove to be more useful for studying germ line imprinting and pluripotency than for performing gene targeting.

The mechanism of EG cells derivation from PGCs also seems to mimic the process of teratoma formation in mice and it will be interesting to see if bFGF and LIF or their cognate receptors are involved in teratocarcinogenesis in mice or humans. It seems likely that the factors described earlier (SLF, LIF, bFGF, and TGFβ1) are not the only factors that regulate PGC development in the mouse embryo. The proliferation rate of PGCs *in vitro* is low compared with that observed in vivo. Although this could be explained by the differences between growth in two or three-dimensional environments, we have probably only begun to scratch the surface in terms of PGC growth regulators. Since the effects of many *W* and *Sl* mutations are not observed until about 9.5 dpc, PGC development up until this name is most likely independent of the *c-kit* signalling pathway. Thus emergence and proliferation of PGCs up until 9.5 dpc might be regulated by a factor other than SLF. A similar pattern is seen during the embryonic development of the hemopoietic system in which many Sl mutations do not have deleterious effects until 13.5 dpc when liver hemopoiesis has commenced. Development of hemopoietic progenitors in the fetal yolk sac is thought to be regulated not by SLF but by another factor.

Tumor necrosis factor-a (TNF-a) is a potent mitoggen for early PGCs (7.5 dpc) in culture and thus could represent one of the factors required for early PGC development. TNF-a was unable to stimulate proliferation of PGCs from 10.5 dpc embryos and thus might act during the *c-kit*/SLF- independent phase of PGC development. As with the other factors identified as PGC mitogens in culture, it will be important to determine if PGCs express receptors for TNF-α. Elucidation of other such factors may be difficult given the small numbers of cells and the problems in isolating them and keeping them alive. One solution to this problem will be to establish immortalized PGC lines. Such cell lines have been produced from postnatal germ cells as well as from PGCs from genital ridges and will undoubtedly be a useful source of both growth factors and growth factor receptors. Ultimately, it would be useful to derive permanent cell lines representative of every stage of PGC development. Somatic cell lines of gonadal origin may also be an important source of PGC growth factors. Such cell lines might provide the appropriate cues to direct normal germ cell development in vitro. The recent demonstration of transmeiotic differentiation of spermatogonia on an immortalized Sertoli cell line is an indication of how useful such cells could be for studying embryonic germ cell development.

Further information about growth factors (or other molecules) involved in PGC development is also likely to come from newly derived "knockout" mice. If such animals survive through any part of the period of PGC development (7.5 to 15.5 dpc), then they can be profitably used to study the role of such factors in PGC development. Two factors that act on PGCs in culture (LIF and TGFβ1) have already been targeted in mice. These animals, and others like them, will be an important resource for studying germ cell development.

Characterization of other mouse mutants that affect PGC development or teratocarcinogenesis may yield important information about PGC growth regulation. Mutations at the *Hertiwgs' macrocytic anemia* (*an*) and *germ cell deficient* (*gcd*) loci affect PGC development, and the timing of action of these mutations suggests that they affect distinct and novel genes. While an is a genetic locus unconnected with *W* and *Sl*, the *gcd* mutation has not been mapped on mouse chromosomes. Similarly, the characterization of gene loci known to affect teratocarcinogenesis, such as *ter*, will undoubtedly lead to a better understanding of PGC development.

Chapter 2

Evolutionary Biology

The variety of life and the ways in which individual types of animals and plants have become adapted in very precise ways to the environments in which they live has excited the interest of natural historians for thousands of years. The turning point in modern biological thought came with the theory of evolution put forward by Charles Darwin and, independently, by his contemporary Alfred Russel Wallace, in the middle of the nineteenth century. Darwin's book presented a mass of observation and reasoned argument supporting 'descent with modification', rather than divine creation of each individual species, as an explanation of the richness and diversity of life.

Darwin provided in the theory of natural selection, a coherent, and in some ways testable, explanation of how and why evolution occurred, invoking only natural biological processes. All species arise by descent from pre-existing ones chiefly by the agency of natural selection acting on the genetic variation normally present within a plant or animal population. Today we take this idea even further back in time than Darwin was able to, to derive the descent of all life on Earth from self-replicating molecules that were generated in the conditions present soon after the Earth was formed some 4600 million years ago. In much simplified terms the original darwinian thesis is as follows. All organisms produce more progeny than can be supported by available resources. This leads to the 'struggle for existence' in which individuals compete with other individuals for scarce resources in the context of prevailing environmental conditions. A variant heritable trait that gives an individual a better chance of surviving to reproductive age and/or producing more surviving offspring will therefore be selected and become more common in the next generation. If it continues to be selected, the character will eventually spread throughout the population.

Selected traits will in the nature of things tend to be adaptive, fitting the organism better for life in its particular environmental niche or enabling it to survive better than its fellows in changing conditions.

Having observed the ability of domestic animals to undergo considerable change and differentiation into distinct breeds under selective breeding, Darwin proposed that natural selection, acting over enormously long periods of time, could give rise to the very great morphological and functional changes that we see in the fossil record and to the differences between present-day organisms. When the theory of natural selection was first proposed the mechanism of inheritance was still quite unknown.

The rapid advances in genetics in the first half of this century led to the neodarwinian synthesis. This was a reinterpretation of Darwin's theory in the light of the newly-discovered principles of genetics, especially of population genetics. Darwinian theory interpreted in its most general sense is still virtually universally accepted as the best explanation so far for the sort of evolutionary change that, for example, produces organisms specifically adapted to a particular way of life or specialized environment.

From time to time the imminent demise of 'darwinism' as an explanation of evolution is announced in the popular scientific literature. But as one leading evolutionary biologist has remarked, reports of its death, like those of Mark Twain, are greatly exaggerated. Running in parallel with mainstream evolutionary biology is the study of evolution at the molecular level, charting the changes in DNA and protein sequences that have occurred as lineages diverged.

Heritable (genetic) change lies at the heart of the evolutionary process, and a record of the changes that have occurred in DNA as generation succeeds generation and diversifies throughout evolutionary time is at least partially preserved in the genes of each of the millions of present-day species.

Like a medieval parchment the record has been continually overwritten and some of it lost for ever, but by piecing together fragments from different sources the history of the genes we carry today can in some cases be reconstructed. The ability to look directly at DNA sequences now makes it possible to trace in some detail the ways in which individual genes and gene families have evolved, and to investigate how complex physiological systems such as the immune system have developed from their primitive beginnings. One important point should be made. In molecular evolution, we are not, in general, dealing with individual genes becoming 'better' over evolutionary time. The differences between the proteins that carry out the same basic metabolic functions in our cells and in those of bacteria are in general trivial. The message coming out of studies of gene evolution is of evolution through the generation of additional new genes by duplication and divergence of existing ones. At present we understand far more about the genes that specify and regulate cellular function and physiology than those that specify morphological structure. Molecular evolutionary studies have, there fore, been confined up to now to genes encoding common enzymes and other well-studied proteins such as the globins and cytochrome *c*. Potentially the most exciting application of molecular biology to evolutionary studies still lies in the future.

The great differences between the various classes of living organism do not lie primarily in the basic biochemistry of their cells, which is constrained by the requirements of life itself. Many enzymes and other essential proteins have remained unchanged in function for thousands of millions of years.

The great differences lie in the way cells have, over evolutionary time, become specialized to perform various functions and become organized into particular structures. In the foreseeable future, the rapid progress now being made in developmental biology and developmental genetics may lie able to provide some clues to the genetic changes that underlie morphological evolution. One practical contribution of molecular biology to general evolutionary studies so far has been to provide a means of assessing the true degree of evolutionary relatedness of one species to another, independent of sometimes subjective and disputed classification on the basis of morphological features.

Comparisons of protein and DNA sequences give, if interpreted with great care, measures of how long ago (in relative terms) the ancestors of two present-day species diverged. Over long periods of time, DNA inevitably accumulates sequence changes by the random process of mutation. If mutations occur at crucial sites that destroy the function of a vital gene, they are usually lost from the population. But those changes that create new and useful genes, or those that occur in 'silent' DNA not in use for coding or gene regulation, will accumulate. As mutation is random, the descendants of two diverging species will accumulate different changes, especially in DNA that is not under selective pressure, and the overall sequence of their DNA will become more and more different over evolutionary time.

The same selective pressure acting on two organisms of different lineages may, of course, produce two rather similar proteins that have evolved independently from quite different ancestral types. Convergent evolution often confuses attempts to trace true lines of descent.

At the molecular level, the difference between similarity by descent and similarity by convergence as a result of selection, is sometimes difficult to determine from protein structure alone. However the nature of most eukaryotic genes, with their extensive regions of non-coding DNA, means that DNA sequences are usually more informative. If the noncoding DNA in two species is still very similar, they must be related by descent and have diverged only recently. The difference between one species and another at the DNA level can be determined in two ways. Suitable corresponding portions of the two genomes can be sequenced and the number of nucleotides at which they differ counted directly. Alternatively the whole of one genome can be matched against the other by DNA-DNA hybridization to get an overall measure of difference.

Along with other biochemical, cytogenetic and molecular studies these techniques have been used to tackle various outstanding questions. A few years ago, for example, cytogenetic and DNA hybridization studies confirmed earlier suggestions that the giant panda *(Ailuropoda melancileuca)* belongs to the bear family, rather than, like the lesser panda (*Ailurus fulgens),* to the racoons. Taxonomists had been divided as to whether the giant panda was a bear, a racoon, or whether it deserved a family all to itself.

DNA sequence and hybridization studies also caused a flutter in some taxonomic dovecotes when they recently placed the chimpanzees as slightly nearer relatives to humans than either of us is to our next nearest relative amongst the great apes—the gorilla. Non-coding DNA is generally used to give unbiased estimates of rates of sequence change, and thus the passage of evolutionary time, within fairly closely related groups of organisms. Charting evolutionary relationships over very long timescales, such as the more than a thousand million years that divide some classes of prokaryotes from each other and from eukaryotes, requires a slightly different approach.

Here one has to look for DNA sequences that evolve only very slowly, as a result of very strong selective constraints. The DNA chosen must evolve sufficiently to provide enough change to distinguish the various lines of descent.

Favoured sequences for this type of work are the rRNA genes, which have been used to trace evolutionary relationships amongst prokaryotes, including the intriguing question of what the last common ancestor shared by eukaryotes and prokaryotes might have been like. The rate of change in nucleotide sequence can in principle be calibrated with fossil evidence to provide a 'molecular clock' counting absolute time elapsed. Molecular clocks presume that over long periods of time the number of mutations (counted as nucleotide substitutions) will be directly proportional to time elapsed (assuming that the deviations from this steady accumulation as a result of selection can be properly taken into account). However, the basic rate of change in the absence of selection turns out to differ widely from class to class for reasons that are so far unknown, but which- may be due to differences in generation times, the fact that some species can repair DNA more efficiently than others, and so on. Sudden large-scale sequence changes may also be introduced by processes other than simple nucleotide substitution which invalidates the basic assumption on which most molecular clocks are calculated. The hope of a universal clock that could be applied to all organisms seems to be receding.

The theory of biological evolution is the mature expression of two revolutionary streams of thought antithetical to a world view that had long prevailed. First, the concept of a changing universe had been replacing the long-unquestioned view of a static world, identical in all essentials to the Creator's perfect creation. People had long sought the causes of phenomena in purposes: the will of God, or Aristotelian final causes (the purposes for which events occur) rather than efficient causes. But Darwin showed that material causes are a sufficient explanation not only for physical phenomena, as Descartes and Newton had shown, but also for biological phenomena with all their seeming evidence of design and purpose. In every scientific discipline, the prevalent ideas and even the questions asked are the products of a historical development. Thus to understand the concerns of modern evolutionary biology, it is essential to known something of the history of the subject. Although the notion of a dynamic world was not foreign to the ancient Greeks, the nonstatic, largely mythological explanation of the origins of living things offered by Empedocles and Anaximander gave way to the philosophy of Plato, which was incorporated into Christian theology and had a permanent and dominant effect on subsequent Western thought. Foremost in Platonic philosophy was the concept of *ειδο** the "form" or "idea," a transcendent ideal form imperfectly imitated by its earthly representatives. The "idea" is an eternal, unchanging essence; thus despite variation among triangles, the sum of the angles of any triangle is 180°, and this "essential" property of triangles distinguishes them absolutely from rectangles. The triangles or horses we see in the material world, according to Plato, are but imperfect copies of the true, perfect Triangle and Horse that exist in the transcendent world of ideas. In this philosophy of essentialism, variation has no meaning; only essences matter. Christian theology adopted an almost literal interpretation of the Bible, including special creation (the direct creation of all things in effectively their present form). But it also incorporated Platonic essentialism in the concept of plenitude.

The eternal, unchanging essences of all things exist in the mind of God, but it would be an imperfection in God to deny material existence to something of which He conceived. Since God is perfect. He must have bestowed existence on everything that existed as His idea. Because to deny existence to anything at any time would introduce imperfection into His creation, all things must have been created in the beginning, and nothing God saw fit to create could have become extinct. Moreover, because order is clearly superior to disorder, God's creations must fit a pattern: the *Scala Naturae*, or Great Chain of Being. This "ladder of life," perceived in the gradation from inanimate material through plants, "lower animals," and humans to angels and other spiritual beings, must be perfect and have no gaps; it must be permanent and unchanging, and every being must have its fixed place according to God's plan. Since this natural order was created by a perfect God, that which is natural is good, and this must be the best of possible worlds. This natural hierarchy extended to higher and lower social classes in human societies. To aspire to change the social order must be immoral, and biological evolution is unthinkable. The role of natural science in this view was to catalogue the links of the Great Chain of Being and to discover their order so that the wisdom of God could be revealed and appreciated. These traditional views gave way under the development of empirical science. Hallowed concepts such as the central position of the earth in the universe were challenged. Newton, Descartes, and others developed strictly mechanistic theories of physical phenomena.

Geologists came to recognize that sedimentary rocks had been laid down at different times, and began to realize that the earth might be very old; it might be as much as 168,000 years old. Fossils that characterized different strata were widely held to reflect a succession of catastrophes such as great floods; some held that there had been numerous successive creations as well. By 1788, however, James Hutton had developed the principle of *uniformitarianism*, which held that the same processes are responsible for both past and present events. This implied that the earth was very old, with no vestige of a beginning-no prospect of an end," as Hutton put it. By the late eighteenth century, the possibility not only of successive special creations, but of the continuing origin of new species by more natural means was bruited about. Maupertuis, Diderot, and Goethe, among others, entertained speculations that new forms of life might originate either by spontaneous generation from inanimate matter, or by the unfolding of immanent potentialities within existing species. "Evolution" in this sense meant not the modification of species, but the manifestation of essences that lay dormant in earlier species.

The earliest uncompromising advocate of evolution was Jean-Baptiste de Lamarck, who first presented an extended exposition of his theory in the *Philosophie*

Zoologique. Lamarck argued not that different living things have descended from common ancestors, but that lowly forms of life arise continually from inanimate matter by spontaneous generation, and progress inevitably toward greater complexity and perfection by "powers conferred by the supreme author of all things"—that is, by an inherent tendency toward complexity.

The particular path of progression taken is guided by the environment, and that a changing environment alters the needs of the organism, to which the organism responds by changing its behaviour, and consequently uses some organs more than others. In other words, use and disuse alter morphology, which is transmitted to subsequent generations. This theory clearly would apply more to animals than to plants.

Lamarck merits respect as the first scientist who fearlessly advocated evolution and attempted to provide a mechanism to explain it. His ideas were almost universally rejected not because he embraced the inheritance of acquired characters, but because leading naturalists of the day found no evidence for evolution.

In particular, Georges Cuvier, the founder of comparative anatomy and one of the most highly respected biologists and paleontologists of the nineteenth century, devastated Lamarck by arguing that the fossil record did not reveal gradual intermediate series of ancestors and descendants, and that organisms are so harmoniously constructed and perfectly adapted that any change would destroy the integrity of their organization. Lyell too, in his *Principles of Geology*, marshalled evidence against evolution in general and Lamarck in particular. In part because of Lamarck, evolution was a common topic of discussion by the middle of the nineteenth century. Robert Chambers's anonymously published

"Vestiges of the Natural History of Creation", for example, was a widely read, rather fanciful evolutionary tract that advanced Lamarckian ideas. But the evidence for evolution had not been marshalled in full, and since Lamarck had been discredited, no satisfactory mechanism of evolution was recognized.

The concept of the "struggle for existence" was used to account for extinction, but no one recognized that it could explain the modification of species except William Wells and Patrick Matthew, who described the concept of natural selection almost as afterthoughts in publications that were devoted to other topics and were little read. The career of Charles Robert Darwin began with his voyage on H.M.S. *Beagle* as ship's naturalist.

Darwin's reading of Malthus's *Essay on the Principle of Population,* which argued that unchecked growth of the human population must lead to famine, may well have been an integral part of his search for a mechanism of evolution, rather than an "amusement" (M. Kottler, personal communication). Twenty years passed between this memorable event and Darwin's first publication on the subject. Perhaps out of fear of the kind of hostility that Lamarck's and Chambers's speculations had met, Darwin occupied himself with acquiring evidence on evolution and worked on a four-volume taxonomic monograph on barnacles that occupied him for eight years. Wallace's theory of natural selection was as carefully reasoned as Darwin's, so he fully deserves credit as the co-discoverer of the chief mechanism of evolution. But although Wallace continued to work on evolutionary topics, especially biogeography, for much of his life, he neither presented a sweeping synthesis as Darwin did in *The Origin of Species* nor explored the ramifications of evolution as Darwin did in his numerous later works.

The Origin of Species has two separate theses: that all organisms have descended with modification from common ancestors, and that the chief agent of modification is the action of natural selection on individual variation. Darwin was the first to marshal on so grand a scale the evidence for the first thesis, the historical reality of evolution, by drawing on all relevant sources of information: the fossil record, the geographic distribution of species, comparative anatomy and embryology, and the modification of domesticated organisms. Much of his argument consists of showing how naturally observations in these areas, such as the vestigial wings of flightless beetles, follow from the supposition of common ancestry, and how implausible they are under the hypothesis of special creation. In developing his arguments, Darwin was among the first to use what has come to be called the *Hypothetico-*deductive method, whereby a hypothesis is tested by determining whether the deductions drawn from it conform to observation. This was not a generally accepted methodology in Darwin's day, when science was supposed to be done by induction (drawing conclusions, as if they were self evident, from an accumulation of individual observations), but it is now generally considered the most powerful method of science. The idea of evolution was not original with Darwin; but the evidence he provided was so strong that it elicited violent reaction from the opposition (especially on religious grounds) and converted all but a handful of prominent scientists to belief in evolution within twenty years. However, his truly original idea, the mechanism of natural selection, convinced few, and indeed fell deeper and deeper into disrepute until the late 1920s. The concept of the "struggle for existence" had been used to explain the extinction of species, but as long as species were viewed as Platonic essences or "types," selection could only eliminate the inferior, not give rise to novelty.

Variation among individual organisms of a species was not mere imperfection, but the material from which selection could fashion better adapted forms of life.

Darwin transformed the Malthusian principle of competition by applying it not only to competition among species, but to competition among individual organisms within a species. Darwin's replacement of essentialism by an emphasis on variation—what Mayr has called "population thinking"—was the foundation of his theory and his most revolutionary contribution to biology.

Evolution generates controversy, like many important concepts, it has been used as a foundation or just as a rationale for philosophical, ethical, or social views.

In the broadest sense, evolution is merely change, and so is all-pervasive; galaxies, languages, and political systems all evolve. Biological evolution is change in the properties of populations of organisms that transcends the lifetime of a single individual. The ontogeny of an individual is not considered evolution; individual organisms do not evolve. The changes in populations that are considered evolutionary are those that are inheritable via the genetic material from one generation to the next. Biological evolution may be slight or substantial; it embraces everything from slight changes in the proportion of different alleles within a population to the successive alterations that led from the earliest protoorganism to snails, bees, giraffes, and dandelions. The several mechanisms of evolution include natural selection, which accounts for the diverse adaptations of organisms to different environments. Neither natural selection nor any of the other mechanisms are providential; natural selection, for example, is merely the superior survival or reproduction of some genetic variants compared to others under whatever environmental conditions happen to prevail at the moment.

Natural selection cannot equip a species to face novel future contingencies, and it has no purpose or goal-not even survival of the species. As environments vary, so do the agents of natural selection-so although trends may be discerned in the evolution of certain groups of organisms, there is no necessary reason to expect a consistent direction in the evolution of any one lineage, much less a direction that all life should follow. Natural selection, moreover, being as purely mechanical as gravity, is neither moral nor immoral. The utterly impersonal mechanistic nature of the mechanisms of evolution seems to be so difficult to grasp-and so repugnant to those who believe that all things exist for a purpose—that meanings have frequently been read into evolution that neither Darwin nor modern evolutionary biologists have intended. Some have equated evolution with "progress" from "lower" to "higher" forms of life, but it is impossible to define any nonarbitrary criteria by which progress can be measured. The very word "progress" implies direction, if not advance toward a goal, but neither direction nor goal are provided by the mechanisms of evolution. Least of all, despite popular conceptions, can evolution be conceived of as being directed toward the emergence of the human species.

The misrepresentation of evolution as progress was so apparent to Darwin that he reminded himself in his notebook "never to say higher or lower" in reference to different forms of life, although he did not always follow his own admonition. One of the legacies of the medieval scholastic principle of plenitude is the "naturalistic fallacy:" the supposition that what is "natural" is "good." Thus "natural laws" are taken not merely as regularities of nature, but become morally binding principles that, as

Collins says, "offer a cosmic backing for the transition from is to *ought*." Evolution and natural selection, like hurricanes and friction, are, but whether they *ought* to be is a question that falls outside the realm of science.

Nevertheless, despite the utter amorality (not immorality) of evolution, natural selection has been taken as a morally proper "law of nature" that ought to guide human behaviour. The "law" of natural selection and the "progress" perceived in evolution were used by Marx to justify class struggle, while the so-called Social Darwinists of the late nineteenth and early twentieth centuries, finding in natural selection an ethical principle that Darwin denied, justified untrammeled economic competition, capitalism, and imperialism.

With more charity but as little logic, Kropotkin and others pointed to the evolution of cooperative social behaviour in animals as justification for more cooperative economic institutions, and the evolutionary biologist Julian Huxley attempted to develop an "evolutionary ethics" that would lead inexorably to higher consciousness and humanitarianism.

Throughout all such thinking runs the supposition that what is natural is good, what is unnatural is bad-a philosophically indefensible posture that persists today when people justify their opposition to birth control or homosexuality by terming them "unnatural." This supposition long predates evolutionary theory, which has merely served as a rationalization for persistent biases.

The two major theses of *The Origin of Species*—that organisms are products of a history of descent with modification from common ancestors, and that the principal mechanism of evolution is the natural selection of hereditary variations—have their counterpart in the two major fields of study that constitute evolutionary biology: the study of evolutionary history and the elucidation of evolutionary mechanisms. The questions, methods, and training of investigators in these two fields tend to be quite different, and rather few biologists have contributed substantially to both fields.

Quite commonly investigators in one of these major areas have been ill-informed on the questions and theories that constitute the other, and have been hampered by their incomplete understanding. The most immediate impact of *The Origin of Species* was to provide a conceptual framework for the study of comparative morphology, descriptive embryology, paleontology, and biogeography, for "relationships" among organisms were now understood to mean common ancestry rather than affinity in the scheme of creation. The data of these studies provided a basis for classification, which was widely adopted as a framework for describing evolutionary affinity. Biogeography and much of paleontology were devoted to inferring the history of spatial distribution and temporal change.

The data of systematics provided, and continue to provide, an immense amount of information on evolutionary trends, patterns of adaptation, the kinds of evolutionary transformation that organisms' features undergo, the intermediate stages in evolutionary sequences, and patterns of variation within species.

The concept of the species itself, partly because of studies in systematics, has been transformed. Many naive ideas, such as Ernst Haeckel's theory that ontogenetic change repeats phylogenetic transformation, have been revised. Although many systematists have devoted themselves chiefly to classification and only incidentally to discovering the evolutionary history of the organisms they study, their work has continued to provide data from which such historical inferences can be made. Systematics has experienced a renaissance in the last twenty years or so, chiefly through the development of more explicit methods for deducing phylogenetic history from such data. Very recently the data of systematics have been greatly enriched by molecular biology; we can now compare not only the phenotypes of organisms, but the sequences of their DNA.

The study of evolutionary mechanisms has had a somewhat more turbulent history. During Darwin's lifetime, the hypothesis of natural selection was under stood by few, and accepted by still fewer. Moreover, the nature of inheritance was not understood.

The observation that offspring are generally intermediate between their parents (in characters such as size) was the basis of a widespread belief in *blending inheritance*, which may be likened to the mixing of two paints or dyes. If blending inheritance obtains, a population will quickly become homogeneous, so natural selection will have no effect; and newly arisen variations will similarly be lost by homogenization. At the same time, the belief that environmentally induced variations could be inherited was widespread, and provided an alternative to natural selection that Darwin himself gave more credence to in his later years. In 1883, August Weismann, an ardent advocate of natural selection, proposed that the germ plasm is entirely separate from and immune to any influences from the soma (the rest of the body), and vigorously rejected any influence of the environment on heredity. Weismann's ideas were strongly attacked by the "neo-Lamarckians" of the period, but were widely accepted after the recognition of Mendel's work. Nevertheless, Lamarckian ideas had their advocates well into the twentieth century, and they have not entirely disappeared today. The discovery in 1900 of Gregor Mendel's demonstration of particulate inheritance should have led to immediate acceptance of Darwin's theory of natural selection.

Given Mendel's theory, the "problem" of blending inheritance held no force; recombination among loci can amplify variation, and new stable variants can arise by mutation. Instead, Mendelian genetics was initially interpreted as a death blow to Darwin's theory.

To oversimplify a complex history, Hugo deVries, William Bateson, and other early Mendelians dismissed continuous variation among individuals as inconsequential and largely nongenetic, and emphasized the role of discontinuous variants that displayed Mendelian ratios and clearly particulate inheritance. Because-they took "species" to be forms that differ discretely in morphology, they believed that species arise in one or a few steps as discrete mutations. If species can arise purely by mutation, their origin does not require natural selection. Thus Darwin's key principles of natural selection and gradual change were dismissed. In the early twentieth century, then, the Darwinian theory of evolutionary change was at a nadir; it was rejected not only by Mendelian geneticists, but by many paleontologists, who espoused "orthogenetic," or directional theories that relied either on intrinsic goal-directed or perfecting principles, or on the assumption that the genetic constitution of organisms constrains them to evolve only in certain directions. Many of the emphases in evolutionary theory since about 1930—the great importance of natural selection, the nature of speciation as the gradual acquisition of reproductive isolation among populations, and the gradual evolution of higher taxa—cannot be understood except in light of the disrepute into which Darwinism had fallen.

Modern evolutionary theory has its foundation in the *Evolutionary synthesis or modern synthesis* that from about 1936 to 1947 forged the contributions of genetics, systematics, and paleontology into a new Neo-Darwinian theory that reconciled Darwin's theory with the facts of genetics. Among the information that contributed to this development were the decisive demonstrations by geneticists that acquired characters-are not inherited, and that continuous variation has precisely the same Mendelian basis as discontinuous variation, entailing the segregation of numerous particulate genes, each with small phenotypic effect.

The variation within and among geographic races had a genetic basis, and that some geographic variations are adaptive. In the late 1920s, Sergei Chetverikov and his coworkers in Russia began to reveal extensive hidden genetic variation in natural populations of *Drosophila,* a program of research that Theodosius Dobzhansky extended greatly after he moved from Russia to the United States in 1927.

Systematists developed the understanding that species are not morphological types, but variable populations that are reproductively isolated from other such populations. The Evolutionary Synthesis emerged not from new information so much as from new concepts.

The theory of population genetics, initiated in 1908 by G. Hardy and W. Weinberg's independent proofs of the "Hardy-Weinberg theorem," was adumbrated in 1926 by Chetverikov but developed in full by Ronald A. Fisher and John B. S. Haldane in England and Sewall Wright in the United States. Fisher in *The Genetical Theory of Natural Selection* and Haldane in *The Causes of Evolution* developed fully the mathematical theory of gene frequency change under natural selection, and showed that even slight selective differences could bring about evolutionary change. From 1917 onward, Wright developed a comprehensive genetical theory that embraced not only selection but inbreeding, gene flow, and the effects of chance (random genetic drift). The content and power of these theories were not fully evident to most biologists, but their major conclusions had an impact.

The elements of theoretical population genetics and extensive data on genetic variation and the genetics of species differences were masterfully synthesized in 1937 in one of the most influential books of the period, Dobzhansky's *Genetics and the Origin of Species.* Ernst Mayr, in *Systematics and the Origin of Species,* elucidated the nature of geographic variation and speciation, incorporating many of the genetical principles that Dobzhansky had articulated.

George Gaylord Simpson, in *Tempo and Mode in Evolution* and its successor *The Major Features of Evolution,* likewise drew on Dobzhansky and Wright in showing that paleontological data were fully consistent with neoDarwinian theory. Julian Huxley's *Evolution: The Modern Synthesis,* perhaps the most comprehensive synthesis of genetics and systematics, did much to establish neo-Darwinism in England. The major tenets of the evolutionary synthesis, then, were that populations contain genetic variation that arises by random (i.e., not adaptively directed) mutation and recombination; that populations evolve by changes in gene frequency brought about by random genetic drift, gene flow, and especially natural selection; that most adaptive genetic variants have individually slight phenotypic effects so that phenotypic changes are gradual; that diversification comes about by speciation, which ordinarily entails the gradual evolution of reproductive isolation among populations; and that these processes, continued for sufficiently long, give rise to changes of such great magnitude as to warrant the designation of higher taxonomic levels.

Although the Modern Synthesis brought about a broad concensus on these tenets, there was still room for disagreement on certain issues. These were chiefly quantitative in nature: Do populations contain a great deal of genetic variation, or is the rate of evolution limited by the rate at which favourable variations arise? Since the Modern Synthesis, the study of evolutionary mechanisms has expanded to incorporate new information, new questions, and new controversies.

The elucidation of the molecular basis of heredity since 1953, when Watson and Crick proposed the structure of DNA, has provided a deeper understanding of the nature of mutation and genetic variation, and has increasingly revealed new phenomena that have enriched and sometimes challenged neo-Darwinian theory. Molecular and other data, as well as an expansion of mathematical theory, have established random genetic drift as a major agent of evolutionary change along with natural selection. The concept of natural selection itself has expanded to include not only differential survival and reproduction of individual organisms, but also, at least in principle, of genes, groups of relatives, populations, and species.

Much of neo-Darwinian theory, including its formalization in the mathematical models of population genetics, is highly abstract; the allele frequencies and selection coefficients that appear in the models apply to traits in general, not to particular features of morphology, physiology, or behaviour. The real features of organisms have increasingly been brought into evolutionary theory by the development of specific models of life history characteristics, particular aspects of behaviour, modes of reproduction, and the like. Entire fields such as behaviour and ecology have become incorporated into evolutionary biology. As this has happened, it has become increasingly clear that the real and present features of organisms cannot be understood solely in terms of existing genetic variations and selection pressures. Rather, present features are in part determined by the developmental processes that translate genotypes into phenotypes, and these in turn are the products of evolutionary history.

The study of evolutionary mechanisms cannot be divorced from the study of developmental biology and of history, the subject matter of systematics and paleontology. The synthesis of these several areas of evolutionary biology has only begun. Evolutionary biology consists of two principal endeavours: inferring the history of evolution and elucidating its mechanisms. In the study of evolutionary history, as in any historical study, inferences about past events are made from typically incomplete and often misleading data.

Sometimes there are historical records (i.e., fossils); quite often there are not, and past events must be inferred from present patterns. Some past evolutionary events, such as the genealogical relationships among species or higher taxa, are determined by the methods of systematics. More recent events, such as those determining patterns of genetic variation within and among populations, can often be inferred from the fit between data and the predictions of population genetics. The historical existence of an event can often be inferred with some confidence, its causes are often much more difficult to elucidate, and frequently must remain a matter for speculation.

From the evidence of comparative morphology and paleontology, we are confident that flowering plants arose from gymnosperms, but we do not know whether the great diversification among flowering plants was caused by their insect pollination, their protected seeds, or their versatile vegetative architecture. That is, the causes of singular events such as the diversification of angiosperms or the origin of World War I are extremely difficult to discern, even in principle. Causes of repeated events, however, such as the frequent evolution of succulence in desert plants, may be inferred if they are correlated with particular conditions. The use of such correlations, the comparative method, has been a powerful tool in evolutionary biology. All sciences progress by formulating and reformulating models and testing them by observation and experiment.

A model may be expressed verbally, as when we postulate that isolation of two populations will permit them to diverge in genetic composition because they do not interbreed; or it may be mathematical, as when we formulate the degree of interbreeding that may permit divergence to occur. The mathematical regularity of Mendelian inheritance is the basis of a genetical theory of the mechanisms of evolution that is probably the most comprehensive and intricate body of mathematical theory in biology. All models, however, no matter how mathematically forbidding, are deliberately oversimplified descriptions of nature. They are not meant to describe the complexities of reality, but to abstract the most general and important features of a process. Because the factors operating in any individual circumstance are far more numerous and complex than a general model can include, the models do not predict future events in detail; the equations of population genetics do not predict the detailed evolutionary path of a population any more than the equations of physics can predict the weather a year from now.

The role of models in evolutionary biology is to specify the conditions under which conceivable events are likely or unlikely to occur, and so to provide a restricted range of possible explanations for observations. Mathematically formulated models also stipulate the quantitative relationships among interacting forces that will permit one event or another to occur.

For example, a simple model will show that a mutation that is deleterious when heterozygous cannot increase in frequency in a large population, even if the mutation is highly advantageous when homozygous. A more complex model shows that such a mutation may indeed increase in a small population, and the smaller the population, the more deleterious the mutation may be and still have an appreciable probability of increase. Models are extremely important in evolutionary theory, but empirical study is obviously important as well. First, observation and experiment are required to describe the phenomena that require explanation-to pose questions. Second, only empirical studies can tell us which of the conditions that a theoretician can conceive of actually occur,, and how frequently: we can conceive of mutations that are deleterious when heterozygous and advantageous when homozygous, but only observation Can tell us if such mutations exist and are common.

Observation can tell us which of the innumerable complexities that might exist are important, and which can be ignored. Moreover, theory will often provide several competing explanations for an observation; empirical study is then required to decide which explanation is most likely. Observation, finally, has another function besides providing support to our intellectual pursuit of understanding the laws of nature: it is the vehicle by which we achieve the esthetic satisfaction of knowing nature. Some biologists have purely intellectual motivations for studying orchids, butterflies, or bacteria; some also find beauty in the pollination of orchids or the structure of a cell.

A few words must be said about "the theory of evolution," which most people take to mean the proposition that organisms have evolved from common ancestors.

In everyday speech, "theory" often means a hypothesis or even a mere speculation. But in science, "theory" means "a statement of what are held to be the general laws, principles, or causes of something known or observed," as the *Oxford English Dictionary* defines it. The theory of evolution is a body of interconnected statements about natural selection and the other processes that are thought to cause evolution, just as the atomic theory of chemistry and the Newtonian theory of mechanics are bodies of statements that describe causes of chemical and physical phenomena. The statement that organisms have descended with modifications from common ancestors—the historical reality of evolution—is not a theory.

Chapter 3

Embryology and Genetics

The specification of embryonic polarity is initiated at early stages of zebrafish development, leading to the establishment of a dorsal blastula signalling center, an equivalent of the Nieuwkoop center, by the midblastula stage. Subsequent rapid and synchronous cellular divisions occur in a stereotyped fashion.

The planes of the first divisions, however, are not correlated with the future dorsoventral axis of the embryo. The initial cleavages are meroblastic, so that all early blastomeres maintain large cytoplasmic connections with a thin cytoplasmic layer of the yolk cell. Starting at the 16-cell stage, only the marginal blastomeres remain connected with the yolk cell.

The synchrony of the early divisions is lost at the midblastula stage, which marks initiation of the zygotic transcription and cell motility. At or soon before the midblastula transition, the marginal blastomeres fuse completely with the yolk cell to form the yolk syncytial layer (YSL). Concurrently, the most superficial blastomeres become an enveloping layer (EVL).

The remainder of the blastomeres located between the EVL and YSL, so-called deep cells, will give rise to all embryonic structures. This characteristic three- layered blastula of zebrafish appears radially symmetric, without any obvious dorsoventral asymmetry.

Dorsal maternal determinants in teleost embryos are thought to be present in the vegetal mass of the yolk cell soon after fertilization. Studies demonstrate that the specification of future dorsal structures is dependent on microtubular arrays present in the yolk cell, and most likely on microtubule-dependent transport of particles from the vegetal pole to the prospective dorsal marginal blastomeres.

The zebrafish zygote transiently exhibits a dense array of parallel microtubules at the vegetal pole, whereas microtubules at the equator do not exhibit any clear organization. During cleavage stages, microtubules extend from blastomeres into the yolk cytoplasmic layer. Disruption of microtubule arrays in the zygote or during cleavage, prior to the 32-cell stage, interferes with animalward movement of particles from the vegetal hemisphere into marginal blastomeres. These treatments inhibit the subsequent axis formation, as judged by the lack of expression of dorsal specific markers at the blastula stage (nuclear β-catenin) and gastrula stage *(gsc, axial).* The resulting embryos either do not exhibit any dorsal axis or possess very reduced axes with medially fused somites lacking notochord and head structures. Furthermore, in medaka fish, the direction of movement of cytoplasmic parcels on cortical microtubules anticipates the future dorsal-ventral axis of the embryo. These studies form the basis for an attractive hypothesis that the asymmetric transport of dorsal determinants from the vegetal pole to marginal blastomeres is one of the initial events involved in the establishment of the dorsoventral axis in teleost embryos. Notably, other work indicates that in amphibians a similar microtubule dependent transport is involved in specification of the dorsoventral axis. In a frog embryo, the dorsal-ventral axis is specified during the first cell cycle, when the cortex rotates relative to the cytoplasmic core along parallel microtubules associated with the vegetal core. Disruption of microtubules blocks the rotation and results in embryos lacking dorsal structures. On the basis of these observations, it has been thought that the rotation of the cortex relative to the core is effecting the translocation of dorsalizing components and dorsal axis induction. However, whereas the dorsalizing components are transported approximately 90° from the vegetal pole, the cortex rotates only 30° during cortical rotation. This apparent contradiction was clarified by the observation that endogenous organelles as well as fluorescent carboxylated beads injected into the vegetal pole are transported along the subcortical microtubules at least 60° toward the future dorsal side. Hence, rather than transporting the dorsal determinants directly, the cortical rotation might serve to align the subcortical microbules.

The identities of substances transported on microtubule tracks are mostly unknown. A notable exception is b-catenin, a component of the Wnt signalling pathway, which was shown to colocalize with subcortical microtubules on the dorsal side of the frog egg at the end of cortical rotation. Furthermore, an important conserved step in the pathway of dorsal axis formation appears to be a nuclear accumulation of b-catenin on the dorsal sides of both *Xenopus* and zebrafish blastulae. This region is thought to constitute the dorsal blastula organizer, which is responsible for the subsequent induction of the dorsal gastrula (Spemann) organizer in the overlying cells. Peter Nieuwkoop first demonstrated the importance of vegetal cells (presumptive endoderm) in induction and patterning of the mesoderm in it series of recombination experiments involving fragments of amphibian blastulae. In those experiments, mesodermal tissues, notochord, muscles, kidney, and blood, formed only when vegetal (endodermal) cells were combined with animal cap (future ectoderm) fragments, but not when vegetal or animal cap fragments were cultured alone. Moreover, the dorsoventral polarity of the induced mesoderm was determined by the particular subregion of endoderm used. Specifically, dorsal mesoderm was obtained only when dorsovegetal fragments were the inducer.

The dorsal mesoderm-inducing activity localized to the dorsal most vegetal cells of the *Xenopus* blastula was named the Nieuwkoop center. β-Catenin is a multifunctional protein involved in cadherm-dependent cell adhesion as well as in the transduction of receptor-mediated intercellular signals β-*Catenin is* a homologue of the Drosophila segment polarity gene *armadillo (arm).* The activity of arm as a signalling molecule is dependent on its nuclear localization, which is regulated by the *wingless (wg)* signalling pathway. In this pathway, the secreted Wg glycoprotein (Wnt in vertebrates) binds to its cell surface receptor thought to be encoded by the *D-frizzled-2* (*fz*) gene.

The ligand-receptor complex stimulates, in an unknown fashion, the activity of a phosphoprotein Dishevelled (Dsh). Activated Dsh in turn inhibits the activity of zeste white shaggy-3 kinase (the homologue of glycogen synthase kinase-3 in vertebrates), which normally destabilizes and thus inhibits arm (β-catenin) activity. In this manner, Wingless signalling leads to nuclear accumulation of β-catenin. In the nucleus, β-catenin interacts with an architectural transcription factor pangolin (LEF-1 family of proteins in vertebrates). In the vertebrate blastula, this complex is thought to activate the transcription of genes involved in axial specification. The likely central role of the Wnt-like signalling pathway in dorsal axis formation in frog embryos is underscored by the ability of several components of the pathway, *dsh,* dominant negative gsk-3 to induce the formation of secondary axes when expressed ectopically in the embryo.

Furthermore, treatment of blastula stage embryos with lithium, which is thought to inhibit GSK-3 kinase activity, also leads to the induction of dorsal mesoderm in the entire marginal zone. However, only P-catenin and GSK-3 have been demonstrated to be required for the formation of the endogenous axis in frog embryos. Notably, dominant negative mutant forms of Wnt and Dsh, while able to inhibit the formation of ectopic axes by their wildtype counterparts, do not prevent formation of the endogenous axis. Therefore, either a redundant pathway or a distinct pathway(s) leading to dorsoventral differences in GSK-3 activity and activation of β-catenin during normal development might exist. An involvement of the Writ signalling pathway components, other than a β-catenin and GSK-3, in the zebrafish embryogenesis remains to be investigated.

Important conjugation experiments in frog embryos indicate that cells in a dorsal vegetal mass, expressing β-catenin, send a "dorsal signal" to other cells at the midblastula stage, after the onset of zygotic transcription. These observations are consistent with β-catenin regulating the inducing activity of the Nieuwkoop center. A systematic analysis of changes in intracellular distribution of β-catenin during *Xenopus* development demonstrated that β-catenin displays greater cytoplasmic accumulation on the future dorsal side of *Xenopus* embryo by the two-cell stage. This is likely to be effected by the dorsally directed microtubular transport discussed previously. The accumulation of β-catenin in dorsal nuclei was reported as early as the 16-to 32-cell stages. In another report, nuclear accumulation of β-catenin was first observed shortly after stage 8 and was most prominent at stage 8.5, disappearing before the blastopore lip formation at the onset of gastrulation. The domain of nuclei positive for β-catenin is located in the dorsal region of the blastula occupying approximately one-third of the equator. Interestingly, this domain extends from the marginal zone into the vegetal and animal regions and is several cell layers deep.

If the distribution of β-catenin positive nuclei reflects the extent of the Nieuwkoop center, it would mean that this organizing center is not located exclusively in vegetal blastomeres but also in marginal and some animal blastomeres.

Alternatively, nuclear accumulation of P-catenin might extend beyond the Nieuwkoop center. In the latter case, activities of molecules other than β-catenin need to be implicated in the specification of the Nieuwkoop center. Where is the Nieuwkoop center equivalent located in a teleost embryo? This question was addressed by transplantation experiments in *Salmo* and in zebrafish, as well as by blastoderm isolation experiments in *Fundulus*. Results of these experiments revealed the key role of the YSL in the process of mesoderm induction and patterning in teleost embryos.

In the blastoderm isolation experiments, relatively normal development occurred in blastoderms deprived of their yolk cell at the 32-cell and later cleavage stages. In contrast, blastoderms isolated earlier did not gastrulate. Oppenheimer hypothesized that the blastoderms removed from the yolk before the 32-cell stage lacked some of the substance that is passed from periblast to blastoderm during or later than the 16-cell stage. It is intriguing that treatments that inhibit the microtubular transport of particles from the vegetal hemisphere to marginal blastomeres, and thus consequently impair formation of the dorsal axis, also are effective only at or before the 32-cell stage. These studies support the notion that the microtubules translocate the "axis-inducing" activity from the vegetal pole to the marginal blastomeres by the 32-cell stage.

However, yolk cells from midblastula stage embryos, when transplanted to the animal pole of another zebrafish embryo, still exhibited the full potential to induce and pattern the host mesoderm. Thus, the axis-inducing activity is present in the yolk cell at the midblastula stage. In zebrafish the YSL forms when the marginal blastomeres collapse onto the yolk cell at about the ninth or tenth cleavage.

Hence, the "axis-inducing" material translocated from the vegetal pole to the marginal blastomeres during early cleavage most likely becomes incorporated into the YSL during a fusion of the marginal blastomeres with the yolk cell. Consistent with till's view, β-catenin-positive nuclei are detected within the YSL on the prospective dorsal side of the high blastula stage zebrafish embryos about 20-30 min after the formation of the syncytial layer. Subsequently, at the sphere stage nuclear localization of β-catenin is observed in the blastoderm cells overlying dorsal YSL. The localization of β-catenin to the dorsal YSL, together with the ability of the yolk cell to induce dorsal mesoderm upon transplantation, strongly suggests that the dorsal YSL in the zebrafish blastula corresponds to the Nieuwkoop center located in the vegetal dorsal blastomeres of the frog blastula. It is noteworthy that the accumulation of β-catenin in the YSL shortly precedes the onset of zygotic transcription at the midblastula transition. Therefore, the initial specification of the dorsoventral polarity and establishment of the inductive center in the dorsal YSL are dependent on the maternal program. A key question is what are the downstream targets of the b-catenin-LEF-I complex in the nuclei of the Nieuwkoop center? Moreover, what is the molecular nature of the "dorsal signal" that establishes the dorsal gastrula organizer in the overlying blastomeres? One potential candidate that might function in both frog and fish embryos is Vg1. This secreted, TGF-b-related molecule was described originally in frogs as maternal mRNA localized to the vegetal blastomeres.

Ectopic expression of mRNA encoding Vg1 fails to induce axis formation. However, injections of mRNA encoding a chimeric BMP-2-Vg1 fusion protein that allows the production of the mature form of the Vg1 ligand in vivo can both restore axis formation in UV-ventralized embryos and induce ectopic axes. In addition, the blastomeres expressing BVg1 appear to act as a Nieuwkoop center. Interestingly, the zebrafish *vgl* orthologue *(zDVR-1)* mRNA is found to be distributed uniformly in the egg and later in all embryonic cells. *zDVR*-1 mRNA fails to induce axis formation when overexpressed in fish embryo. However, the normal zDVR-1 precursor does appear to be processed to mature protein when expressed in *Xenopus* and can act as a potent inducer of axial mesoderm in this system. These observations form the basis for a proposal that localized posttranslational processing of Vgl precursor protein on the prospective dorsal side plays a regulatory role in the development of the dorsal axis in frog and fish embryos.

How is the proposed activation of Vg1 related to the β-catenin pathway? Because mature Vg1 ligand has not been detected during normal development, the timing of its presumably proteolytic activation *in vivo is* not clear. However, *BVgl* mRNA injections can rescue the effects of depleting maternally encoded β-catenin in frog embryos, suggesting that *VgI* acts downstream of β-catenin and thus around the midblastula transition. These experiments cannot, however, exclude the possibility that *VgI* acts in a pathway parallel to β-catenin. Indeed, additional evidence from *Xenopus* indicates that *Vgl* acts synergistically with a maternal Wnt-like signal to specify dorsal fates in both mesoderm and endoderm. An important finding has been that a downstream target of β-catenin in Xenopus embryos appears to be the siamois gene which encodes a paired-class homeodomain protein. Injections of RNA encoding Siamois into from embryos lead to the induction of complete secondary axes, but the progeny of injected cells to not participate in the secondary axes. Siamois expressed after the midblastula transition and is present most abundantly in the dorsal endoderm of early gastrulae, indicating that it might be an important component of the Nieuwkoop center. Interestingly, whereas β-catenin and other components of the Wnt pathway highly activate expression of members of the *siamois,* TGF-P pathway (such as VgI) were shown to activate siamois only weakly. The involvement of TGF-P signalling in the formation of the zebrafish gastrula organizer is further supported by the activity of nodal and *TARAM-A* genes.

The murine *nodal* gene is a TGF-P-related ligand that is essential for the formation of the node, the mouse equivalent of the Spemann organizer. Injections of mouse nodal mRNA into zebrafish embryos lead to ectopic expression of the organizer-specific genes *gsc* and *liml* and formation of ectopic axes containing notochord and somites. TARAM-A is a serine threonine kinase type I receptor related to TGU-β and activin receptors. Whereas ectopic expression of RNA encoding wild-type TARAM-A leads only to an expansion of the axial mesoderm, injections of RNA encoding a constitutively active receptor kinase, TARAM-A-D, result in a massive induction of the dorsal and panmesodemial markers, gsc and ntl shortly after the midblastula stage as well as numerous dorsal mesoderm genes in the late blastula stage (50% epiboly). Furthermore, injections of RNA induce the formation of complete secondary axes in which cells that inherited injected RNA mostly contribute to the anterior dorsal mesoderm (hatching, and head mesoderm). Notably, TARAM-A mRNA is found to be distributed uniformly in the zygote and during cleavage stages and disappears at the 500-cell stage from the marginal blaslomeres that are fated to form the YSL. In the mid-and late blastula. TARAM-A mRNA is detected at the blastoderm margin around the circumference of the embryo, with its expression increasing on the dorsal side starting at 40% epiboly. Both the activity and the expression pattern of TARAM-A make it a good candidate for a transducer of signals from the Nieuwkoop center involved in formation of the Spemann organizer.

Another link between the Nieuwkoop center, P-catenin, TGF-P pathway, and the Spemann organizer in zebrafish has been revealed by the zebrafish zygotic *bozozok* mutation. Homozygous *boz*^{m168} mutants are characterized by the lack of the main derivatives of the gastrula organizer, prechordal plate and notochord, and also exhibit defects in the neuroectoderm.

Notably, an embryonic shield cannot be detected morphologically in *boz* mutants at the early gastrula stage, and expression of all tested organizer-specific genes is reduced or absent, indicating that *boz* function is required for the proper formation of the dorsal gastrula organizer.

The *boz* phenotype is reminiscent of defects observed in embryos in which microtubules were disrupted during cleavage stages, a treatment that prevented nuclear accumulation of β-catenin and inhibited dorsal axis formation. Several additional observations indicate that boz acts during the blastula stage within or downstream of the Nieuwkoop center in the pathway specifying axial mesoderm.

First, expression of organizer-specific genes is already reduced at the late blastula stage. Second, whereas ectopic expression of β-catenin leads to the induction of full secondary axes in wild-type embryos, the axes induced in *boz* embryos lack notochord and other midline cell types. Thus, β-catenin is unable to suppress the *boz* phenotype.

The lack of notochord and prechordal plate, the hallmarks of the *boz* phenotype, is fully suppressed by ectopically expressed mRNAs encoding mouse nodal and the activated form of TARAM-A-D, but not by wild-type TARAM-A. These studies indicate that the *boz* locus is required in a TGF-β signalling pathway that establishes the dorsal gastrula organizer in zebrafish acting downstream or in parallel with dorsalizing signals like β-catenin. Transplantation experiments performed in the 1920s by Hilde Mangold and Hans Spemann identified the dorsal blastopore lip of the amphibian gastrula as a tissue that, upon transplantation to the ventral side of the gastrula, will induce formation of the secondary body axis.

The organizer tissue, while contributing mostly to chordamesoderm and to a lesser extent to paraxial mesoderm and the floor plate of the neural tube of' the secondary axes, was able to induce and pattern the ectopic neuraxis as well as coordinate gastrulation movements. All vertebrates exhibit an equivalent dorsal gastrula organizer region, as defined by its inductive properties and expression of an array of organizer-specific genes, including gsc, *HNF-3β, Xlim-1, Xnot, noggin, chordin, siamois, ADMP,* and others. In zebrafish, the embryonic shield, a dorsal thickening along the blastoderm margin of the early gastrula, is thought to correspond to the Spemann organizer. During normal development, this region gives rise to axial mesoderm, notochord and prechordal plate, and to some neural tissue, predominantly floor plate.

Transplantations of the shield to the ventral blastoderm margin lead to the formation of ectopic axes, with the donor tissue forming notochord, prechordal plate, and floor plate of the neural tube. Interestingly, the ectopic axes resulting from the shield transplantations are not complete and usually lack the anterior head structures.

One explanation is that in these experiments the extirpated and/or transplanted tissue did not contain the entire organizer. This could be due to experimental difficulties with extracting the entire shield, leading to residual organizer activity remaining in place. Indeed, the expression of organizer-specific genes was not monitored in the operated embryos. It is noteworthy that a different technique, glass pipet versus eyelash knife, can be used to perform the shield extirpation-transplantation experiments, allowing one to remove most of the region expressing organizer-specific genes, such as *gsc* and *flh*.

The shield extirpations performed with a glass capillary result in embryos with the axial mesoderm reduced or missing completely and severe deficiencies in the anterior neural structures, including cyclopia and/or eye reduction. However, neural keel still forms in these experimental embryos, indicating that a significant level of organizer activity might reside beyond the shield region in the zebrafish embryo. Alternatively, some inductive events happen before the shield stage. Consistent with this notion, the expression of a number of genes like *gsc*, *lim 1*, and *flh* can be detected on the dorsal side of the embryo before the embryonic shield manifests itself. Support for the notion that the organizer in fact extends beyond the shield comes from the observation that one of the key organizer-specific *genes*, *chordino* (zebrafish chordin homologue), is expressed in a broad dorsal marginal domain extending outside the shield.

The dorsal gastrula organizer is thought to carry out a number of inductive and patterning functions: induction and patterning of the neuroectoderm, dorsalization of paraxial mesoderm, coordination of gastrulation movements, and finally self-differentiation into midline structures.

Discoveries in a number of animal systems, most notably expression cloning in *Xenopus*, identified several molecules that might fulfill these distinct functions of the gastrula organizer. Identification of numerous mutations affecting organizer function in zebrafish provides genetic formation of the presumed functions for some of these molecules, as well as identification of novel genetic components. Furthermore, the analysis of the zebrafish mutants reveals genetic hierarchies that govern pattern formation in the gastrula.

Numerous molecular biological and genetic studies support the notion that the two main activities of the dorsal gastrula organizer, dorsalization of mesoderm and neural induction (which could be viewed as dorsalization of ectoderm), are accomplished by secreted factors that antagonize ventralizing signals derived from the ventral organizing center in the gastrula. This interaction between bone morphogenetic proteins 2 and 4, secreted ligands of the transforming growth factor-b superfamily, and their antagonists appears to be a mechanism for embryonic dorsoventral patterning that is conserved from the fruit fly to vertebrates. The realization that the formation of ventral fates in the embryo is [lot default pathway, but rather depends on specific signalling events, was suggested initially by experiments showing that ectopic expression of BMP-4 induced ventral and posterior cell types in animal cap explants and whole *Xenopus* embryos. Transcripts encoding BMP-4 are present in *Xenopus* unfertilized eggs and initially are distributed uniformly throughout frog blastula. In contrast to *Xenopus*, transcripts for the zebrafish homologues are not found in the maternal pool of RNAs. *z-bmp*-2 transcripts are detected first at the sphere stage and are distributed uniformly throughout the blastoderm. However, as the yolk donies at the onset of epiboly, *zbmp*-2 transcripts disappear from the presumptive dorsal side. At 50% epiboly, just before the onset of the involution/ingression movements that will create the germ layers, *zbmp-4* transcripts become detectable. Similar to the *Xenopus* embryo, at the early (shield) stages of gastrulation, both *zbmp*-2 and -4 exhibit high levels of transcripts in the presumptive ventrolateral blastoderm in a domain extending from the margin up through the animal pole and ire absent on the dorsal side. An interesting and not yet fully understood exception is it small distinct domain of *zbmp*-4 expression in the shield region itself, marking the precursors of the anterior axial mesoderm.

Overexpression of *zbmp*-2 and *zbmp-4* in zebrafish embryos supports the conclusions drawn from studies of frog embryos that these factors respecify dorsal mesoderm to ventral mesoderm and expand the prospective epidermis at the expense of neuroectoderm. However, the exact phenotypes reported differ in an interesting way. Neave and colleagues performed the most systematic analysis by injecting varying amounts of *Xenopus bmp*-4. RNA (20-220pg/ embryo) into 1-4-cell-stage zebrafish embryos. Progressively ventralized embryos were observed in a dose-dependent manner. The lowest *bmp-4 dose* led to "ventralized I" class embryos, which lacked head and notochord structures, whereas the somites lost their chevron shape and became fused in the midline. In the "ventralized II" class of embryos, besides the lack of a head and notochord, the yolk extension failed to form.

More ventralized embryos were observed only after the injection of 100 pg or more RNA. Embryos of class III exhibited radial symmetry; cells accumulated at two opposite poles of the round yolk cell with somitic tissue forming at one of the ends. Finally, embryos of class IV likewise were radially symmetric; however, cells accumulated only at one end, suggesting that epiboly also was affected. A similar loss of axial structures and often severe reduction of the hypoblast was observed in embryos injected with 100 pg of *zbmp-2* RNA. Analysis of axial mesoderm markers *gsc* and *ntl* revealed that expression of these genes was normal in *bmp-4* RNA-injected embryos at the onset of gastrulation and that reduction or absence of gsc and *ntl* expression began to be observed by 65-70% epiboly. These results indicate that even high doses of BMP-2/4 signalling do not prevent the initial formation of the main component of the dorsal gastrula organizer, axial mesoderm.

However, exclusion of BMP-2/4 signalling from the dorsal side of the gastrula might be important for maintenance and further development of axial mesoderm, as has been reported earlier for *Xenopus*. Interestingly, a distinct phenotype was caused by ectopically expressing BMP-4 at later stages of development by using a plasmid with *bmp-4* cDNA under the control of the cytoskeletal actin (CSKA) promoter from *Xenopus borealis,* which directs BMP-4 expression only after the midblastula transition. Whereas there was expansion of ventral markers and reduction of neuroectodermal markers in injected embryos during gastrulation, at the end of embryogenesis only the posterior part of the notochord was reduced and embryos exhibited multiple ventral fin folds and an enlarged blood island. It is possible that the phenotype obtained with DNA injections represents the mildest ventralization and thus could reflect the lowest dose of

BMP-4 signalling obtained with injections of 20 pg of RNA, as reported by Hammerschmidt *et. al.* Over-expression of antagonists of BMP-2/4 signalling in zebrafish embryos leads to a reduction of ventral fates and to an expansion of dorsal fates including neural tissue as reported previously for frog embryos. Injections of synthetic mRNA encoding a truncated, dominant negative form of the *Xenopus* BMP receptor (tBR) led to ectopic domains of expression of the prechordal plate marker, the *gsc* gene, whereas overexpression of *Xenopus noggin* mRNA expanded the dorsal expression domain of gsc. Overexpression of both types of BMP antagonists resulted in enlargement of the notochordal rudiment as judged by the midline expression domain of *ntl* or axial genes. The expansion of axial mesoderm was associated with a reduction of intermediate mesoderm, as monitored by expression of the *pax2* gene in the kidney primordia, as well as with the reduction of ventral mesodermal markers such as *gatal.*

Within the ectoderm, ectopic expression of both *MR* and *noggin* mRNAs resulted in the expansion of neuroectoderm precursors expressing *fkd3* at the expense of nonneural ectoderm, as marked by reduced expression of *gta3, dlx3,* and *eve l.* Importantly, the ventral but not dorsoanterior expression domain of BMP4 also was reduced or absent in these dorsalized embryos. This is consistent with positive feedback loop regulation of *bmp*-4 expression in the ventral region of the zebrafish embryo, which has been postulated previously in *Xenopus.* A consistent picture of dorsoventral patterning in the vertebrate gastrula emerges from the preceding observations. A concentration gradient of a BMP-2/4 morphogen (as either homo-or heterodimer) provides an instructive signal determining the range of dorsoventral fates in the mesoderm and ectoderm. This ventrodorsal gradient of BMP-2/4 signalling arises as a combination of the expression of BMP-2/4 genes, which is maintained by a positive feedback loop, and the expression of BMP antaconists that bind to the BMP ligands and block signalling through their receptors.

The preceding model is now supported by analysis of mutations from large-scale genetic screens for mutations affecting zebrafish embryonic development. Mutations defining at least 15 genes affect patterning during gastrulation. An initial characterization of mutant phenotypes led to the classification of these mutations into two groups, based on their effects on the formation of cell fates in the embryo. One class of mutants exhibits a loss of cell fates derived from the dorsal part of the gastrula, including neuroectoderm, whereas the second class of mutants has deficiencies in ventral and posterior structures, in some cases accompanied by an expansion of neuroectoderm..

Within the first class, mutations in six loci, including cyclops (cyc), *floating head, bozozok (boz), one-eyed pinhead (oep),* and *schmal*spur, result predominantly in deficiencies of mesodermal and neuroectodermal cell fates derived from 'the dorsal region of the gastrula (axial mesoderm and anterior- ventral neuroectoderm), without an obvious increase in ventrally derived fates. Mutations in three loci, chordino *(din)*, and *ogon, mercedes (mes)* lead to an increase in ventral fates and concomitant decrease dorsolateral fates, mostly somites and neuroectoderm. The phenotypes of mutants in each of the preceding classes can be strikingly phenocopied either by ectopic expression of BMP-2/4 from RNA or DNA constructs or by injections of RNAs encoding BMP-2/4 antagonists, as described earlier. Specifically, the phenotypes resulting from mutations at *din, mes,* and *ogo* loci are characterized by multiple ventral fin folds, increased numbers of blood cells, and increased expression levels of ventral markers like *evel*. Expansion of ventral fates is accompanied by a reduction of neuroëctoderm, somitic, and chordameso-derm in *din* mutants.

All of the preceding defects also were observed in embryos after injections with *bmp-4* cDNA on a CSKA plasmid. In addition, the phenotype of the *din* mutation can be suppressed by ectopic expression of *tBR* and *noggin,* further supporting the notion that the defect in this mutant is in the inhibition of BMP-2/4 signalling. On the other hand, the phenotypes of the "dorsalized" class of mutants at the *Swirl* (*swr*), *somitabun* (*sbn*), *snailhouse* (*snh*), *lost-a-fin* (*laf*), *piggytail* (*pgy*), and *mini fin (mfn)* closely resemble defects obtained by ectopic expression of the BMP-2/4 antagonist.

Moreover analysis of *swrdin* double mutants demonstrated that swr is epistatic to *din*. These observations gave support to the hypothesis that *din* encode an inhibitor of swr function, with *din* and *swr* affecting the components of the morphogenetic system itself, *chordin* and *BMP-2/4*, respectively. It is very satisfying that this hypothesis has been confirmed by evidence that *din* alleles affect the zebrafish *chordin* homologue. First, *din* mutant allele and the zebrafish *chordin* gene were shown to be closely linked (<0.25 cM). Second, the din^{tt250} mutation was demonstrated to introduce a deletion of 104 base pairs in the *chordino* ORF that led to a frame shift. The resulting, truncated $din^{tt\text{-}250}$ mutant protein had only the first 42 amino acids of the wild-type form, thereby lacking the conserved cysteine-rich repeats. Third, injections of *Xenopus chordin* mRNA into din^{tt250} mutants led to a full rescue of the mutant phenotype. These observations provide genetic evidence for the requirement of Chordin function as a BMP-2/4 antagonist in the Spemann organizer of a vertebrate embryo. Notably, two groups have demonstrated that, as hypothesized, the *swr* locus encodes the morphogen itself, BMP-2. Single strand conformation polymorphism (SSCP) was used to show linkage between *zbmp-2* and the *swr* locus. Furthermore, cloning and sequencing of the two swr alleles, swr^{172} and swr^{AC100}, revealed that each exhibited a point mutation in the *zbmp-2* ORF. Upon RNA injection, the swr^{AC100} mRNA was demonstrated to be inactive. Interestingly, mRNA encoded by the swr^{A72} mutant allele was weakly dorsalizing, suggesting that Swr^{A72} might be an antimorph.

In addition, injections of wild-type *zbmp-2* RNA can rescue the *swirl* mutant phenotype. Embryos from a cross of a rescued *swirl* homozygous female with a wild-type male appear completely normal, indicating that *zbmp-2* has no strict materrial function. This is consistent with the lack of maternal stores in the *zbmp-2* mRNA in zebrafish eggs. It will be crucial to determine whether the remaining mutations from the "dorsalized" group affect known components of the BMP-2/4 signalling pathway, like type I and type II receptors or MAD proteins, or whether they will identify new components so far undetected by nongenetic analyses. Interestingly, the phenotype of boz mutants—lack of a notochord and prechordal plate as well as deficiencies in the ventral neural fates—is phenocopied by injections of low doses of BMP-2/4 mRNAs. Furthermore, these aspects of the boz mutant phenotype also are phenocopied by inhibition of microtubules at early cleavage stages as well as by shield extirpations, as discussed previously. Therefore, this class of phenotype might identify processes and genes involved in the specification, development, and maintenance of axial mesoderm in zebrafish. Indeed, there are differences in the timing at which development of axial mesoderm is affected in these different experimental situations. In *boz* mutants, expression of axial mesoderm markers is affected even before the onset of gastrulation, as early as 40% epiboly. In contrast, in embryos injected with mRNA encoding BMP-4, expression of gsc is normal at the beginning of gastrulation, becomes reduced, and finally disappears by midgastrulation. Therefore, whereas inhibition of microtubules at early cleavage stages and the *boz* mutation both seem to affect specification of axial mesoderm or its maintenance before the onset of gastrulation, overexpression of *bmp-4* RNA affects the maintenance of axial mesoderm derivatives (notochord and prechordal plate) during gastrulation.

The initial expression of prechordal plate markers and their later disappearance or reduction in the course of gastrulation also is observed in *cyc,* and *sur* mutants. Therefore, these genes might be required for the maintenance or further development of prechordal plate derivatives. It is intriguing that injections of plasmid encoding BMP-4 as well as the mutations in the *din* locus, lead only to gaps in the posterior notochord or do not affect notochord formation. Furthermore, *boz* mutants do not exhibit a very characteristic feature of *din, nies,* and ogo mutants: multiple caudal fin folds. Nevertheless, could both classes of phenotypes, *boz* and *din,* result from ectopic BMP-2/4 signalling caused by decreased inhibition of BMP-4 action? One possibility is that the two phenotypes reflect different amounts of BMP-4 signalling, with the *din* phenotype representing low levels and the boz-like phenotype higher levels of ectopic BMP-4 activity. Another possibility is that the phenotypic differences observed in these various experimental situations are not caused solely by different levels of BMP-4 signalling, but rather by the differences in the timing of the BMP-4 ectopic activity. Indeed, injections of RNA would lead to an earlier accumulation of BMP ligands, even in early blastula, than in the case of plasmid injections, which would bring about expression of the, protein only after the midblastula transition. It is also tempting to speculate that not only would the onset of ectopic activity be different in these two cases, but so would its duration.

If *chordino* functions during late gastrulation and somitogenesis, it is likely that the multiple ventral fins might reflect the ectopic activity of BMP-4 at late stages of gastrulation or even during segmentation stages. Indeed, BMP-4 is expressed at the edges of developing caudal fins during segmentation.

Hence, it is likely that its ectopic activity at these stages might affect fin development. In contrast, shield extirpations and *boz* mutations might not affect BMP-2/4 activity at these later stages of development. It will be very informative to compare carefully the timing and levels of expression of *zbmp-2/4* and *chordino* between these different phenotypes. Our understanding of the mechanisms underlying the anterior-posterior (AP) patterning of the CNS is still very rudimentary. In all existing models of AP neural patterning, the involuting axial mesoderm is a key player. According to one model, distinct regions of axial mesoderm directly induce specific regions of the CNS along the AP axis of the overlying ectoderm. Alternatively, AP patterning is a two-step process, with all of the initially induced neuroectoderm being specified as forebrain by a so-called activator.

The more posterior neural structures, midbrain, hindbrain, and spinal cord, then result from the anterior neuroectoderm being influenced by a second signal, a transformer. Other experiments have demonstrated that fragments of the lateral margin of zebrafish gastrula posteriorized forebrain when transplanted into the animal pole region, resulting in ectopic expression of hindbrain markers. Therefore, the transformer signals might be emanating from the lateral blastoderm margin of the zebrafish gastrula, which indeed underlies future hindbrain progenitors. Noggin and Chordin are thought to induce neural tissue of anterior character, thus acting as activators in the two-signal model of Nieuwkoop. Upon the influence of transformer signals, including Xwnt3A, retinoic acid, and fibroblast growth factor (FGF), more posterior neural fates would arise. How does this model hold up in view of the phenotypes obtained with overexpression of BMP-2/4 and its antagonists and mutations affecting the dorsoventral pattern in zebrafish embryos? Consistent with observations from frog embryos, overexpression of BMP-2/4 in zebrafish, or the increased BMP-2/4 activity in the *din* mutants, leads to dorsal expansion of nonneuronal ectoderm at the expense of neural tissue.

An opposite effect is observed in experiments in which embryonic BMP-2/4 signalling was compromised by ectopic expression of BMP antagonists, as well as in dorsalized mutants like.

Indeed, in *swr* mutants, neural tissue seems to be present around the entire circumference of the embryo, as evidenced by a dramatic circular expression of the *pax*2 midbrain marker.

A similar phenomenon is observed for *fkd3* in swirl mutants, as well as in *noggin*-and tBR-injected wild-type embryos. What is the dorsoventral (DV) and anterior-posterior patterning of the diminished and expanded neural tube in "ventralized" and "dorsalized-mutants respectively? In embryos injected with BMP-4 encoding mRNA, the decreased neural tube shows a reduction of ventral fates, cells expressing *shh* as well as primary motoneurons. However, the loss of ventral neural fates in these embryos might be a secondary defect due to axial mesoderm reduction or loss that takes place during gastrulation.

Although the neural plate of *din*^{c4} mutants is reduced, these mutants possess axial mesoderm and they also exhibit ventral neural fates like floor plate and primary motoneurons. This would support the notion that the absence of these cell types observed after the injection of *bmp-4* RNA is a secondary effect due to the loss of midline mesoderm rather than a direct influence of dorsal expansion of BMP-4 activity.

Interestingly, the *din* mutants were not reported to exhibit increased numbers of dorsal-specific sensory neurons. Rather, the overall number of primary neurons was decreased. The underlying cause of this difference between *din*^{c4} mutants and embryos injected with *bmp-4* RNA needs to be investigated. Still, little is known about the AP pattern in the reduced neural plates of ventralized" mutants and their induced phenocopies. *din*^{c4} mutants show reduced expression of the anterior neural marker *otx2* at late stages of gastrulation (80% epiboly) and reduction of *otxl* expression during the segmentation period (20 somites).

Hindbrain expression of *Krox20* is present during early segmentation (4-6 somites). However, the expression domains of *Krox20* in rhombomeres 3 and 5 appear reduced in the mediolateral and AP directions. The proportional decrease of *otxl*, *otx2*, and *Krox20* expression domains suggests a mediolateral reduction of neural plate along the entire AP axis, while the AP pattern remains relatively normal. Similarly, in 22-somite-stage *din* mutants all expression domains of *pax2* are present along the AP axis of the neural tube, such as prospective optic stalks, midbrain, and otic vesicle anlagen. In contrast, the *boz* mutants show not only a more narrow neural plate but also a clear abnormality in its AP neural pattern. At day 1 of development, severe reduction of the forebrain and midbrain regions and lack of eyes are observed. The nature of defects in the AP patterning is revealed by expression of a *sekl* gene. In wild-type embryos, *sekl* is expressed in the forebrain region and in rhombomeres 1, 3, and 5 of the hindbrain. Remarkably, in the *boz* mutant embryos the forebrain expression domains are missing. This is accompanied by a longitudinal expansion along the AP axis of the three rhombomeric domains of *sekl* expression. Another remarkable property of the dorsal gastrula organizer is its ability to induce and orchestrate gastrulation movements characteristic of the dorsal side of the gastrula. Similar to other vertebrate embryos, in teleosts, the main components of the organizer, prospective prechordal and chordamesoderm involute/ ingress and move toward the future head region below the prospective neuroectoderm.

Furthermore, cells of mesendoderm and ectoderm precursors converge toward the organizer. While in the dorsal region of the gastrula, these cells as well as ectodermal cells participate in mediolateral intercalation, effectively leading to extension of the embryonic body along the anterior-posterior axis. These convergent extension and involution/ingression movements are preceded and accompanied by epibolic movements during which all the cell types of the zebrafish embryo, YSL, all germ layers, and the enveloping layer expand vegetally to cover completely the yolk cell by 10 hpf.

It is noteworthy that practically every single cell of the zebrafish gastrula is engaged simultaneously in these three types of movements, epiboly, involution/ ingression, and convergent extension, as well as in inductive interactions that pattern germ layers. Identification of the cues that these cells use to navigate through the embryo and make decisions about their prospective fates presents a formidable challenge. Gastrulation mutants identified in the genetic screens, as well as experiments with expression of BMP-2/4 and its antagonists, reveal important links between the inductive events in the embryo and gastrulation movements. Furthermore, the mutant phenotypes uncover interdependencies or the lack thereof between distinct morphogenetic movements.

The first important lesson we learn from the mutant phenotypes is that the process of epiboly appears to be independent of DV patterning. Indeed, epiboly occurs relatively normally in all the mutants discussed previously in which DV defects are observed.

Also, the reverse is true: mutations affecting epiboly do not lead to obvious defects in the DV pattern of the gastrula. Furthermore, epiboly is not affected significantly in embryos that were dorsalized or ventralized by various experimental treatments like injections of b-catenin, TARAM-A-D, BMP-2/4, and BMP-2/4 antagonists. The lack of a relationship between epibolic movements and DV patterning is not that surprising, considering that epiboly is initiated at the sphere stage just before the first asymmetric expression of dorsal-specific genes like gsc can be detected and that it occurs around the entire circumference of the embryo.

However, there is a certain asymmetry to the rate of epibolic expansion of blastoderm along the. DV axis, with the blastodem spreading across the yolk slightly faster on the dorsal side of the embryo. As a consequence, the point at which the yolk plug is closed is not at the vegetal pole but slightly ventral to it. This asymmetry can be explained, though, by convergent extension movements that are concentrated on the dorsal side where they contribute to epibolic spreading, of the blastoderm, creating the DV difference in the rate of epiboly. It will be important to investigate whether this asymmetry is decreased or eliminated in mutants in which convergent extension is impaired.

Involution/ingression movements at or near the blastoderm margin lead to the formation of a mesendodermal hypoblast below the ectodermal epiblast in the teleost gastrula. This movement occurs around the entire circumference of the embryo, although it might be initiated first and might be more intense on the dorsal side of the gastrula.

None of the mutants identified so far exhibits a very obvious defect in hypoblast formation, but more subtle defects cannot be excluded at this point. However, strongly affected embryos injected with BMP-2/4 RNA exhibit a clear reduction of hypoblast at the tailbud stage that might be a consequence of impairment of involution/ingression movements.

Convergent extension is an important morphogenetic movement of gastrulation in all vertebrate embryos that leads to the narrowing of an embryo in the mediolateral axis and its AP elongation. In teleosts, convergence of cells from ventral and lateral positions toward the dorsal side of the gastrula leads to thickening of the embryonic shield and formation of the embryonic body rudiment. Cells converging dorsally leave the ventral, and later the lateral, area, creating an evacuation zone that increases with the progress of convergent extension. Mediolateral cell intercalations that are thought to occur mostly in the dorsal region of the gastrula drive both further narrowing and extension of the embryonic axis and of an rudiments in the AP direction. Not only the rudiment of the embryonic body but also the shape of the entire gastrula undergoes characteristic changes, most likely effected by these convergent extension movements.

The embryo changes its shape from a sphere at the onset of morphogenetic movements to an oval (elongated in the animal-vegetal axis) at the end of epiboly.

Mutations defining at least 16 genes result in an abnormal shape of the gastrula that could be explained by abnormal convergent extension movements. At the bud stage, *kny, tri, slb, ogo, din,* and *oep* mutant embryos exhibit a distinctly shorter embryonic axis, indicative of decreased AP extension. All of these mutants also are characterized by a spherical yolk cell at the tailbud stage, in contrast to the yolk cell of wild-type embryos, which is elongated in the animal-vegetal axis at this stage of development. This is most likely a consequence of a decreased convergent extension movements. Indeed, the evacuation zone is smaller in *kny* and *tri* mutants than in the wild type at this stage of development, supporting the notion that convergence extension movements are decreased in these embryos. Expression domains of several genes in the ectoderm and mesoderm are perturbed in the mutant embryos in a manner consistent with decreased convergence and extension; the expression domains are expanded mediolaterally and compressed along the AP axis of the embryo. Embryos with an abnormally elongated "squash" shape are observed in the *swirl, somitabun (sbn), snailhouse (snh), piggytail (pgy), lost-a-fin (laf), and mini fin (mfn)* mutants and for some alleles of *chrodino,* as well as embryos injected with RNAs encoding β-catenin, Noggin, and tBR. This phenotype appears to be opposite that observed for mutants with decreased convergent extension. Surprisingly, the "Squash" shape of swr embryos does not reflect a simple increase in convergent extension movements. Rather, the dorsal convergence of cells is decreased as judged by the thickened ventral anterior region of *swr* mutant embryos at 80% epiboly. Furthermore, the notochord of these embryos is broader, possibly a consequence of reduced convergence. How can a reduction of convergence lead, on the one hand, to shorter and rounder embryos such as kny, tri, and slb, but on the other hand to the excessively elongated gastrula in the swirl-like mutants? Is it possible that the two classes of phenotypes arise due to different defects in the second type of cellular behaviour thought to drive convergent extension, i.e., mediolateral intercalation.

The rounder shape of the gastrula, the shorter but broader embryonic axis of the *kny* and *tri* class of mutants, might reflect a reduction in mediolateral intercalations concomitant to reduced dorsal convergence.

In swirl-like mutants, cells would exhibit excessive mediolateral intercalation behaviour, not only in the dorsal region of the gastrula but also around the circumference of the embryo. This would be exaggerated by the presence of cells on the ventral side of the embryo due to reduced convergence. Furthermore, the phenotype of din and ogo mutants might result from reduced rates of mediolateral intercalation without convergence being affected significantly. Testing this hypothesis requires precise in vivo analysis of cell movements in these mutants by time-lapse video-microscopy. Notably, very interesting parallels exist between the two types of convergent extension defects and patterning of the gastrula, *din* and *ogo* mutants from the "ventralized" class exhibit a rounder shape of the gastrula and shorter body axis, whereas the "dorsalized" class of mutants, like swr, is characterized by the elongated shape of the gastrula. Furthermore, embryos injected with RNA encoding Noggin or tBR also are characterized by the elongated shape of the gastrula. The dorsalization of the gastrula caused by these manipulations might be the underlying cause of convergent extension defects in such embryos. First, dorsal convergence is likely to be directed by a signal from the shield. Furthermore, in *Fundulus embryos* the net rate at which cells move toward the dorsal side of the gastrula is inversely proportional to their distance from the shield. Hence, the signal for convergence might be graded, with a peak at the dorsal side. It is also conceivable that cells might be actively moving away from the ventral side. Because in "dorsalized" embryos expression of dorsal-specific genes is expanded toward the ventral side, so might be the expression of the signal for dorsal convergence. In extreme cases, such a signal might be distributed uniformly around the circumference of the embryo, resulting in inhibition of convergence movements. Similarly, the ventral expansion of a signal for mediolateral intercalation behaviour might result in ectopic mediolateral intercalations in the ventral part of the embryo, as discussed earlier.

Indeed, in Xenopus embryos ectopic expression of Noggin can induce, either directly or indirectly, elongation of tissue explants. It has been demonstrated that narrowing and extension of the dorsal marginal zone explants is affected by mediolateral intercalation of cells in the explant. In "ventralized" mutants the signals for mediolateral intercalation would be decreased, and this could explain both the shape of the gastrula and the decreased extension of the axis. As mentioned earlier, the convergence defect is not obvious in these mutants.

Chapter 4

Sex Determination

Sex determination and dosage compensation are among the best understood developmental pathways due to the insights provided by many laboratories. The chapter opens by addressing the nature of sexual dimorphism in the nematode C. elegans and the primary determinant of sex—the X:A ratio.

Early events in dosage compensation and sex determination are regulated coordinately by a set of early acting genes, but this coordinated control resolves into separate pathways with independent sets of genes responsible for executing each process. Until the complexities associated with the coordinated control of sex determination and dosage compensation were appreciated, progress in analysing the X:A ratio was delayed.

The nature of the XA ratio and the strategies for identifying counting elements Lire revisited after a discussion of this coordinated control. The review then turns to the topic of germ-line sex determination and addresses the nature of germ-line-specific regulatory mechanisms that allow the self-fertile hermaphrodite to produce sperm first and then switch to producing oocytes.

The small nematode *C. elegans* has two naturally occurring sexes: male and self-fertile hermaphrodite. The hermaphrodite has a female soma, but her germ line first produces a limited number of sperm before making oocytes. Otherwise, she can be considered the female of the species. One consequence of the production of both sperm and oocytes in an essentially female animal is that the mechanisms controlling somatic and germ-line sex determination are not identical, although many of the same global genes are involved. The specification of germ cells to become sperm requires the repression of female-promoting genes and the activation of male-promoting genes. Subsequently, the male-promoting genes must be repressed to allow the switch to oogenesis. In *C.elegans*, the decision to follow the hermaphrodite or male fate gives rise to differences in body patterning, behaviour, and biochemistry. The lineage of each cell in the hermaphrodite and in the male is known in its entirely.

Most of the differences in sexual phenotype arise postembryonically, however, the first sign of sexual dimorphism appears during embryo-genesis and is marked by two sets of programmed cell deaths. These deaths remove four CEM (male-specific cephalic neurons) in XX hermaphrodites and two HSN (hermaphrodite-specific neurons required for egg-laying) in XO males. Over 30% of somatic nuclei in the adult hermaphrodite and 40% in the adult male are sexually dimorphic; the most obvious differences are manifested in tissues involved in reproduction and mating.

The XX hermaphrodite has a bilobed ovotestis that reflexes to meet centrally at the vulva (opening for egg-laying on the ventral surface), whereas the male has a single-lobed gonad that produces only sperm. The hermaphrodite also has neurons and muscles that are involved in egg-laying. The tail of the hermaphrodite tapers to a simple spike, whereas the male has an elaborate tail with specialized features and sensory structures that are necessary for mating and copulation. Males also display a characteristic mating behaviour, and it has been possible to identify genetic mutants that disrupt this behaviour. These include many mab mutants, which have abnormal male tail morphology, and many cod (for copulation defective) mutants, which have no obvious physical defects in male tail morphology, yet fail to display normal mating behaviour.

At the biochemical level, yolk proteins (vitellogenins) are expressed exclusively by the intestine of XX hermaphrodites. Ectopic expression of yolk proteins by the intestine of XO males has proven to be a sensitive reporter for inappropriate XO feminization. Males, although not essential for the propagation of wild-type *C. elegans*, are found at a frequency of 0.2% of the natural population. *C. elegans* XO males do not have a Y chromosome and arise through meiotic non-disjunction. Although wild-type XX hermaphrodites are self-fertile, they can be mated with XO males to produce cross-progeny. Sperm provided by XO males have a competitive advantage over hermaphrodite sperm, so that after mating only cross-progeny are produced in a Mendeliain sex ratio of 1:1.

It was shown by Nigon that wild-type diploid XX animals are hermaphrodites and XO animals are males. What is the primary determinant of sex in *C. elegans*? Is it the absolute number of X chromosomes? To address this question, Madl and Herman varied chromosomal ploidy to determine how it affected sex determination. Normally, diploid XO animals (X:A ratio, = 1X:2A = 0.5) are male and diploid XX animals (2X:2A = 1) are hermaphrodite. Their results indicated that the primary determinant of sex is the X:A ratio (number of X chromosomes to the number of sets of autosomes) and not the absolute number of X chromosomes. By way of example, XX tetraploids (2X:4A = 0.5) develop as males, not hermaphrodites, which would have been the case if the absolute number of X chromosomes determined sex. Further examination to different X:A ratios revealed that animals with an X:A ratio of' 0.67 (2X:3A) developed as males and those with an X:A ratio of 0.75 (3X:4A) developed as hermaphrodites. One consequence of the dimorphism in X chromosome number between the sexes is that X-linked expression must be equalised through dosage compensation. Several dosage compensation schemes have been documented in other organisms.

In mammals, dosage compensations is achieved through X inactivation. X inactivation does not appear to be the basis for *C. elegans* dosage compensation, because XX animals heterozygous for an X-linked recessive marker invariably are wild type in phenotype. Instead, measurements of the steady-state mRNA levels of several X-linked *C. elegans* genes indicate that dosage compensation is achieved by downregulating, by one-half, the expression of each X chromosome in an XX hermaphrodite so that it is equivalent to the Output from the single X in XO males. Variations in ploidy that lead to severe overexpression (4X:2A) or underexpression (1X:3A) of the X chromosome result in lethality or severe sickness. These excessive imbalances in X chromosome dosage probably cannot be redressed by the dosage compensation machinery. Similarly, mutations ill genes controlling dosage compensation might also be expected to produce sex-specific lethality in diploids. A set of five autosomal genes (dpy-21, -26, -27, -28, and -30) were identified that implement X chromosome compensation.

It is generally observed that Mutations in genes controlling dosage compensation are lethal to XX animals and have little or no effect on XO animals. Strong loss of function or null mutations in either *dpy-26*, -27, or -28 lead to XX inviability and rare dumpy (Dpy, short and fat) escapers, which give rise to the dpy gene name; in contrast, mutations in these same genes have essentially no effect on XO males with a single X chromosome. *dpy-21* mutants show less severe, but similar defects; XX dpy-21 hermaphrodites are Dpy and XO males essentially are wild type. XX *dpy-30* alleles are temperature sensitive and their phenotype is somewhat different from other dosage compensation dpy mutants are almost always inviable, but at permissive temperatures rare Dpy escapers are observed; XO dpy-30 animals are viable males, but scrawny, indicating that dpy-30 also plays a role in XO development. Mutations in the dosage compensation dpy genes have no direct effect on sexual phenotype, although the implementation of dosage compensation is determined by the X:A ratio. Therefore, the dosage compensation process is distinct from the pathway controlling sex determination. It has been observed, however, that mutations in the dosage compensation dpy genes can indirectly influence sex determination. 2X:3A (X:A = 0.67) animals that would normally develop as males can be shifted toward the hermaphrodite mode of sex determination by mutation in the dosage compensation dpy-21 or *dpy-27* gene.

Mutations in the dosage compensation dpy genes have been shown to elevate X-linked expression in XX animals by using phenotypic assays and by measuring X-linked expression, consistent with the proposed wild-type role of these genes in downregulating X-linked expression. Because of the similar mutant phenotype shared by most dosage compensation dpy genes, they have not been ordered by genetic means, but it is postulated that the dpy gene products function together to implement dosage compensation. Not all X-linked genes are dosage-compensated in XX animals; the levels of an unknown X-linked mRNA, *uxt-2*, are 2-fold higher in XX hermaphrodites than in XO males. It is possible tnat the X:A numerator elements also will escape dosage compensation.

Substantial progress has been achieved toward understanding the molecular and biochemical basis of dosage compensation in *C. elegans.*

The Molecular analysis of dpy-27 has provided the first clues that dosage compensation functions to lower X-linked chromosome expression by specifically altering the architecture of the X chromosome. By using antibodies directed against the DPY-27 protein, it was shown that, prior to the 30-cell stage, DPY-27 protein is present in the nuclei of both XX and XO animals; however, at the 30-cell stage, DPY-27 associates with each X chromosome in XX hermaphrodites, but fails to localize to the single X in XO males. Thus, the onset of dosage compensation is associated with the 30-cell stage. The dpy-27 gene encodes a homologue of the SMC family of proteins, which play essential roles in chromosome condensation and segregation in yeast and frogs. Ill XO vol-1 mutants, DPY-27 is inappropriately associated with tile single X chromosome; therefore, XOL-I normally prevents DPY-27 from associating with the single X chromosome XO animals. The dosage compensation gene, *dpy*-26, was cloned and found to encode a novel acidic protein. Similar to DPY-27, antibodies directed against DPY-26 protein detect DPY-26 in the nuclei of both XX and XO animals; at the 30-cell stage, DPY-26 is associated with each X chromosome in XX, but not XO animals. DPY-26 also co-localizes with DPY-27 to the X chromosome at each cell cycle, however, the staining patterns of DPY-26 and DPY-27 are not identical. Unlike DPY-27, DPY-26 protein is associated with all condensed chromosomes as they Undergo mitosis prior to the 30-cell stage. However, by the 60-cell stage, DPY-26 staining is limited to the mitotic X chromosome in XX animals. Genetic analyses indicate that, in addition to its role in dosage compensation, *dpy*-26 plays it role in meiotic segregation; dpy-26 mutants display a higher level of X chromosome non-disjunction as detected by the increase in XO males. Examination of the XX hermaphrodite germ line using anti-DPY-26 antibodies shows that DPY-26 co-localizes with germ cell chromosomes as they condense and enter pachytene. In contrast, DPY-27 is not detected in the hermaphrodite germ line.

It has been further shown that mutations that abolish DPY-26 staining in embryos do not affect the localization of DPY-26 protein in the hermaphrodite germ line. Therefore, DPY-26 has an independent function associated with meiotic chromosome segregation in the hermaphrodite germ line that is separable from its role in dosage compensation.

To examine how the dosage compensation dpy genes and their upstream regulators interact to regulate the assembly of a complex on the X chromosome, the effects of mutations in these genes on the immunochemical staining patterns of DPY-26 and DPY-27 were examined. The *sdc* genes are responsible for activating dosage compensation in XX animals, and xol-l is responsible for repressing the sdc genes in XO animals. It was found that *sdc*-2, *sdc*-3, and *dpy*-30 are necessary for localizing DPY-27 to the X chromosome in XX animals. DPY-27 fails to localize to the X chromosome in XX animals mutant for *sdc*-2, *sdc*-3, or *dpy*-30, and the staining pattern instead resembles that XO animals. Similarly, DPY-26 also fails to localize to the X chromosome in *sdc*-2, *sdc*-3, *dpy*-27, and *dpy*-30 mutants. The *dpy*-26 and *dpy*-28 genes affect the synthesis or stability of DPY-27. Mutations in these genes do not significantly alter *dpy*-27 mRNA steady-state levels, but they completely abolish DPY-27 staining. Mutations in *dpy*-27 and *dpy*-28 also diminish or abolish DPY-26 staining, but do not affect *dpy*-26 mRNA steady-state levels. It was also reported that null mutations in *sdc*-1 or *dpy*-21 have no effect on the staining patterns of either DPY-27 or DPY-26, indicating that these genes do not play a direct role in recruiting DPY-27 or DPY-26 to the X chromosome. At the biochemical level, co-immunoprecipitation experiments indicate that DPY-26 and DPY-27 form a complex and that this interaction may be important for protein stability. Additional studies suggest that DPY-28 may also complex with DPY-26 and DPY-27. It remain to be determined how XOL-I regulation of the sdc genes or their products prevents a dosage compensation complex from forming on the single X of XO males.

Perturbations in dosage compensation often lead to sex-specific lethality; however, sexual transformations in either direction generally are not lethal. Thus, early genetic screens led to the recovery of many sex reversal mutants that have phenotypes opposite to that specified by the X:A ratio. These mutants were shown to carry mutations in genes that globally regulate both somatic and germ-line sex determination. Genetic selection and suppressor screens subsequently have made it possible to identify novel genes that display tissue specificity and rare alleles of known sex-determining genes, which have shed light on tissue and sex-specific regulatory mechanisms. The three *tra-1*, -2, and -3 genes were among the first identified loss-of function mutations in genes that regulate sex determination. Mutation in any *tra* gene *(tra-1,2,3)* transforms XX animals into males; however, the phenotypes of these transformed XX males are not identical. XX *tra-1(lf)* males are capable of siring, cross-progeny, whereas XX *tra-2(1f)* or *tra-3(lf)* males are not. XX *tra-2* and *tra-3* males show defective tail morphology and an absence of mating behaviour; hence, these males also are referred to as pseudomales. In contrast to their effect on XX sexual phenotype and behaviour, mutations in *tra*-2 and *tra*-3 do not affect the normal development of XO males; however, *tra-l* XO males have variably abnormal gonads, indicating that *tra-1* plays a role in the normal development of the XO gonad. The haplo-insufficiency associated with *tra-2* is one indication that sexual cell fate decisions are sensitive to perturbations in the dose or activity of the sex-determining genes.

Genes that promote XO male development also were identified through genetic screens by searching for mutations that feminize XO animals. *her-l(lf)* mutations transform XO animals into fertile hermaphrodites, but have no effect on XX animals. Null mutations in the *fem*-l, -2, and -3 genes transform both XO and XX into spermless hermaphrodites, that is, females.

The *ferm* genes are essential for both XX hermaphrodite and XO male spemiatogenesis. Dominant gain-of-function feminizing mutations in the *tra-l* gene also were recovered: both XX and XO *tra-l(gf)* mutants develop as females, which can be mated to produce cross-progeny. Temperature-sensitive alleles were identified for many of the sex-determining genes and dosage compensation *dpy* genes, which have proven to be valuable tools for identifying extragenic suppressors.

Temperature shift studies using temperature-sensitive alleles of sex-determining genes have revealed that the sexual fate of individual tissues is determined at different times in development. It has also been shown that many of the sex-determining genes are required continuously throughout development to maintain a sexual cell fate decision.

For example, *her-1* activity is required continuously in XO males to ensure the maintenance of the male state;

when XO *her-l(ts)* males raised continuously at permissive temperatures are shifted as adults to restrictive temperatures, they produce oocytes and their intestine begins synthesizing yolk proteins. The genes described so far are global regulators that control sexual cell fate decisions in both somatic and germ-line tissues, but have no apparent role in dosage compensation. The control of germ-line sex determination involves additional tissue-specific genes and tissue-specific controls that regulate the activity of global sex-determining genes. Therefore, the following discussion pertains only to the control of somatic sex determination.

Epistatic interactions among the different sex-determining genes have been analysed extensively. For example, XX animals carrying a single mutation in either *tra-2* or *fem-1* develop as males or females, respectively; XX animals doubly mutant for *tra-2* and *fem-1* develop as females, not males. Therefore, in a control hierarchy *fem-1* acts later than *tra-2*. After testing all possible double-mutant and some triple-mutant combinations, the sex determining genes were ordered in a hierarchical regulatory pathway based on negative regulation. In this pathway, sex determination is controlled by a series of on-off or high-low switches, whereby each gene functions to negatively regulate the activity of its nearest downstream neighbour. The *tra-1* gene is epistatic to all sex-determining genes: *tra-1(gf)* double mutants develop as females, not males. Therefore, *tra-1* is the master regulator of somantic sex determination: female somatic development ensues when *tra-1* is inactive.

The genes discussed so far function to promote only dosage compensation or only sex determination. However, additional genes have been identified that coordinate both pathways. The first of these genes to be identified was *sdc-1* (for sex and dosage compensation).

Mutations in *sdc-1* have no effect on XO males, but they produce dumpy and variably masculinized XX hermaphrodites. Thus, *sdc-l* disrupts both dosage compensation and sex determination.

Two lines of evidence support the view that *sdc-1* mutations disrupt dosage compensation: first, X-linked genes are overexpressed in XX sdc-I mutants; second, the Dpy phenotype is associated with XX animals, regardless of their sexual phenotype. It was further shown that loss-of function mutations in her-l, which transform XO animals to hermaphrodites, suppress the masculinization of XX *sdc-1* mutants, but not their inability to activate dosage compensation. This indicates that *sdc-1* independently regulates dosage compensation and sex determination.

On the basis of genetic epistasis arguments, *sdc-1* has been placed upstream of both tile dosage compensation and sex determination pathway *sdc-1* does not function alone to coordinately regulate dosage compensation and sex determination. *sdc-2* and *sdc-3* mutants also exhibit XX sexual transformations and defects in dosage compensation, while having no effect on XO males.

Of the sdc genes, null mutations in *sdc-2* exhibit the most severe phenotype; XX *sdc-2* mutants are inviable, but rare escapers are Dpy and sexually transformed. sdc-3 is unusual ill that, in addition to null mutations that produce Dpy XX hermaphrodites, certain sdc-3 alleles affect only dosage compensation or only sex determination. Thus, *sdc-3* appears to have independently mutable regions that discriminate between its effects on sex determination and dosage compensation. The *sdc* genes are not the earliest genes in the regulatory hierarchy that respond to the X:A ratio; this position presently is occupied by *xol-1*. *xol-1* null mutations have no effect on XX animals, but they are lethal and ferminizing to XO animals.

Ectopic expression of *xol-1* also is lethal to XX animals because the male mode of dosage compensation is activated inappro-priately. The lethality of *xol-1* mutants is suppressed by mutations in any of the dosage compensation *dpy* genes, whereas the sex determination defects are suppressed by mutations in any of the *sdc* genes. Taken together, genetic epistasis suggests that *xol-1* functions upstream of the *sdc* genes and that *xol-1* is a direct target of the X:A ratio.

Therefore, *xol-1* is the earliest acting regulatory switch gene affecting both sexual cell fate and dosage compensation. *vol-1* appears to be transcriptionally regulated because steady-state RNA levels are higher in XO than in XX embryos. In addition, *vol-1* also may be regulated post-transcriptionally, because the *-vol-1* locus expresses multiple transcripts, only one of which is sufficient to rescue all aspects of the *xol-1* mutant phenotype.

The molecular cloning of the major regulatory genes controlling sex determination provided insights into, but not an immediate understanding of, the mechanisms Underlying the genetic pathway of sex determination. Several of the sex determining genes were shown to have sequence motifs found in other proteins with chara-cterized function; however, many of the sex-determining genes encoded novel proteins. Among those proteins with known sequence motifs were *sdc-l*, *sdc-3*, and *tra-l*; each of these genes encodes proteins with zinc finger motifs, which are associated with transcriptional regulation. From this information, it was surmised that transcriptional regulation plays a key role in the initiation and execution of sexual cell fate decisions. However, transcriptional regulation is not the only mechanism controlling sexual cell fate; as discussed in Sections IV B and IV C, cell-to-cell signalling and signal transduction also play central roles in regulating the final activity state of *tra-1*.

What is the transcriptional target of *sdc-1* and *sdc-3?* One potential target is *her-1*. At the beginning of the pathway controlling only sex determination, *her-l* is the earliest acting switch gene *her-l* also is active in XO males and is inactive in XX hermaphrodites. This difference in *her-1* activity arises through transcriptional regulation; *her-1* mRNA steady-state levels are high in XO males and low or absent in XX hermaphrodites. The steady-state levels of *her-1* mRNAs respond to the X:A ratio and are not affected by mutations in downstream sex-determining genes. For example, although XX *tra-1* mutants develop as phenotypic males, the levels of *her-1* mRNAs in XX *tra-1* males are comparable to those found in wild-type XX hermaphrodites but not XO males. The *sdc* genes are good candidates for mediating the transcriptional repression of *her-1*. Consistent with this notion, *her-1* mRNA steady-state levels are 20-fold higher than those of wild type in animals carrying an sdc mutation. As stated earlier, *sdc-1* and *sdc-3* each encode proteins with zinc finger motifs: however, the zinc fingers in SDC-3 have been shown to be necessary for dosage compensation, but not sex determination. Presently, the mechanism by which SDC-3 and the other SDC proteins repress *her-l* is unknown.

A more difficult challenge has been to Understand how the presence or absence of *her-l* is detected and subsequently transduced through *tra-2, tra-3* and the fem genes to control *tra-1* activity. Part of the understanding came from observations indicating that sex determination contains a cell nonautonomous component. Mosaic analysis indicated that *her-1* and *sdc-1* behave cell nonautonomously. Molecular analysis subsequently has revealed that *her-1* encodes a predicted secreted protein called HER-l. Because *sdc-1* regulates the expression of *her-1* mRNAs, this might also explain why *sdc-1* functions cell nonautonomously despite encoding a zinc finger protein that might be expected to function cell autonomously. Another piece of puzzle was provided when it was shown that one of the *tra-2* transcripts encodes a large predicted membrane-spanning protein named TRA-2A which might act as a receptor for the HER-l ligand.

A simple model was proposed that took into account the genetic relationships of the genes controlling sex determination, the cell nonautonomy of sexual cell fate decisions, arid the molecular identities of the sex-determining genes.

The key aspects of this model involved cell-to-cell signalling mediated by HER-1 and TRA-2A and signal transduction mediated through TRA-2A and the intracellular FEM proteins to TRA-1 TRA-2A is depicted as being present in both sexes, because tra-2 mRNAs are present in both sexes, albeit at levels 10-fold lower in XO males. In XO males, HER-1 is present and is postulated to bind to and inactivate TRA-2A. In turn, the FEM proteins are active and repress TRA-1, possibly through protein-protein interactions or post-translational modification. In XX hermaphrodites, HER-1 is absent and TRA-2A is predicted to function constitutively, promoting female development in the hermaphrodite. TRA-2A is postulated to bind to and inactivate one or more of the FEM proteins, possibly through sequestration, allowing TRA-1 to promote female development.

TRA-1 targets

Genetically, tra-l promotes female development either by activating the expression of female-specific genes, such as yolk proteins (vitellogenins), or by repressing the expression of male-specific genes. The *mab-3* gene (for male abnormal) has been the most promising candidate for a downstream target for TRA-1 because of its specific role in promoting male development. Mutations in *mab-3* lead to inappropriate expression of yolk- proteins by the intestine of XO males and to defects in the differentiation of sense organs in the male tail. Surprisingly, *mab-3* encodes a protein with sequence similarity to Drosophila doublesex (dsx), a DNA binding protein that plays a key role in the regulatory pathway of Drosophila sex determination. MAB-3 contains two copies of the DNA binding motif that is found in DSX; this motif has been named DM. Not only do MAB-3 and DSX share structural similarities, but they also appear, to have conserved function.

The *dsx* gene produces two sex-specific alternatively spliced transcripts that encode proteins DSXF and DSXM with alternative carboxy termini. DSXM controls sex-specific neuroblast formation and yolk protein gene expression, similar to the role of MAB-3 in worms. Transgenic expression of DSX^M, but not DSX^F, partially restored the formation of male tail sensory organs in XO mab-3 mutants. The conservation of two proteins involved in controlling sexual cell fate across phyla has been hitherto unprecedented. Additional downstream targets of TRA- 1 may possibly be identified by taking advantage of the *C. elegans* sequence, which will be >95% complete in 1998. The TRA-1 protein has zinc finger motifs that preferentially bind a nonamer sequence, similar to that identified for the human GLI protein; TRA-1 is a homologue of GLI and GL13. Therefore, TRA-1 binding targets would be predicted to carry one or more copies of the TRA-1 binding domain in their promoters. Although a number of candidate genes with such properties have been identified by the *C. elegans* genome sequencing consortium, appropriate tests have not yet been performed to examine whether these genes are regulated by TRA-1.

The demonstration that the X:A ratio is the primary determinant of sex was one of the earliest contributions to our understanding of *C. elegans* sex determination. However, the identification of "counting" elements has been more refractory to analysis.

In partial explanation, duplications or deficiencies harbouring numerator elements may not have been recovered because many of the pre-existing X chromosome duplications and deletions were isolated on the basis of their viability in XX and XO animals. Chromosomal duplications that increase the number of numerator elements would be predicted to cause XO-specific lethality, and similarly, deletions that lower the dose of numerator elements might cause XX-specific lethality. A greater understanding of the coordinated control of sex determination and dosage compensation has led to the development of strategies that allow the recovery of deletions and duplications that might otherwise be detrimental be cause they result in the implementation of an inappropriate mode of dosage compensation.

For example, duplication of X chromosome numerator elements that are feminizing and lethal to XO animals can be suppressed by mutations in an sdc or dosage compensation has led to the development of strategies that allow the recovery of deletions and duplications that might otherwise be detrimental because they result in the implementation of an appropriate mode of dosage compensation. For example, duplication of X chromosome numerator elements that are feminizing and lethal to XO animals can be suppressed by mutations in an sdc or dosage compensation dpy gene or by constitutive expression of *xol-1*. Conversely, deletions that remove numerator elements, and would be predicted to be lethal and masculinizing to XX animals, can be suppressed by an xol-l mutation, which locks animals into the XX mode of dosage compensation and sex determination.

Using the strategies described previously, the left end of the X chromosome was found to harbour potential numerator elements; duplications of this region had detrimental effect on XO viability, which could be suppressed by a mutation in dosage compensation dpy gene. It was further found that the region of the X chromosome harbouring these numerator elements could be subdivided into three regions, each containing, numerator activity. Duplications covering two or more X chromosome subregions revealed that the combination of numerator elements increased XO lethality synergistically. On the other hand, removal of a single dose of all of the left-hand X chromosome numerator elements produced only a weak Sdc phenotype, indicating that XX animals may better tolerate increases in X-linked expression than XO animals can tolerate decreases. Taken together, these studies indicated that numerator elements are likely to be discrete elements and that there are probably only a limited number of these elements.

Molecular identification of one numerator element was aided by comparing two duplications, *eDp26* and *mnDp73*, which differ in XO lethality because *mnDp73* carries only the two weak regions at the left end of the X, whereas *eDp26* carries an additional strong region. The region of the X chromosome unique to *eDp26*, and not *mnDp73*, was delineated by fluorescent in situ hybridization. A small number of overlapping cosmids in this region, when introduced as multicopy transgenes, were found to mimic the XO-specific lethality and feminization produced by *eDp26*, and a likely candidate gene with potential numerator activity was identified, *fox-1* (feminizing on X). The *fox-1* region encodes a putative RNA binding protein, suggesting, that it may function as a transcriptional or post-transcriptional regulator of *xol-1*.

Subsequent genetic screens to identify X-linked suppressors of the XO lethality associated with duplications of the left region of the X chromosome identified two point mutations in the *fox-1* gene. Therefore, *fox-1* mutations function in a dose-dependent manner to suppress the effect of X chromosome duplications that are lethal to XO animals: however, XX animals deleted for the *fox-1* gene are fully viable, indicating that fox-l is not an essential gene and the other signals participate in the counting process. To explore the notion that *fox*-1 directly controls the regulatory switch gene *xol-1*, the expression of an integrated *lacZ* reporter driven by the *vol-1* promoter, *Pxol-1::lacZ*, was examined in *fox-1* mutants. Integrated copies of the *Pxol-I::lacZ* reporter transgene behave appropriately by sex; high levels of reporter activity are detected in XO animals and low levels in XX animals. It was found that FOX-1 does not directly regulating the transcription of *vol-1*, because the *Pxol-l::1acZ* reporter remains repressed in XX for *fox-1* mutants. However, when transgenic animals carrying a bifunctional reporter composed of *xol-1* genomic sequences fused in frame to GFP (green fluorescent protein) were exposed to high levels of transgenic FOX-1, both the reporter and the rescuing activity of the construct were repressed. This result suggests that FOX-1 regulates *xol-1* post-transcriptionally. The previous report that the *xol-1* locus expresses multiple transcripts raises the possibility that FOX-1 regulates the alternative splicing of *xol*-1. Therefore. X chromosome numerator elements appear to function by regulating *vol-1* activity at both the transcriptional and post-transcriptional levels.

At the present time, denominator or autosomal counting elements have not been identified. One might surmise that if the denominator were a single gene, then screens for XO feminizing mutants would have already identified such an element. However, if the denominator is polygenic or is not a chromosomal entity per se (e.g., the nuclear volume occupied by all chromosomes), then simple genetic screens may not easily identify such elements. The global sex-determining genes affect not only somatic sex but also germ-line sex. However, there are several important differences between the pathways controlling sex determination in the soma and in the germ line. First, although *tra*-l is involved in germ-line sex determination, it is not the terminal regulator of cell fate in the germ line. Doubly mutant *fem;tra-1* animals have a male soma, but they produce oocytes, indicating that *tra-l* is not directly required for specifying female fate in the germ line. Second, additional *mog* and fog genes have been identified as tissue-specific regulators of germ cell fate. These genes regulate the specification of oocytes and sperm, respectively, but play no apparent role in somatic sex determination Third, the *fem* genes, *fog-1*, and *fog-3* are the terminal regulators of sex in the germ line. These genes are absolutely required for spermatogenesis to occur in XX and XO animals, and their absence results in the specification of all germ cells as oocytes. Finally, several of the global regulators of sexual cell fate have germ-line-specific control regions that are involved in restricting their activities during the hermaphrodite sperm-oocyte decision.

The hermaphrodite has a bilobed tubular somatic gonad that meets centrally at the vulva. A pool of mitotically proliferating germ cells are maintained at the distal end of the gonad: a somatic cell named the distal tip cell (DTC) plays a key role in maintaining this pool of proliferating germ cells. Laser ablation of the DTC results in all germ cells entering the meiotic pathway; this phenotype is mimicked by mutations in the *glp-1* gene (for germ-line proliferation). Molecular analyses reveal that the maintenance of mitotically proliferating germ cells is dependent on cell-to-cell signalling mediated by the GLP-1 receptor in the germ line and the LAG-2 ligand, which is expressed by the DTC. GLP-l is a transmembrane protein related to the C. *elegans* LIN-12 and *Drosophila* Notch receptors. LAG-2 is a membrane protein with sequence similarity to *Drosophila delta,* which encodes a ligand for Notch.

The molecular analysis of mutant alleles that specifically affect the regulation of *tra-2* and *fem-3* activity in the hermaphrodite germ line indicated that post-transcriptional regulation mediated through 3' untranslated regions (UTRs) and protein-protein interactions are important fit controlling the hermaphrodite sperm-oocyte decision. The two classes of *tra-2* germ-line-feminizing mutants identify two distinct controls that regulate the onset of hermaphrodite spermatooenesis. The six *tra-2(gl)* mutations all disrupt at least one copy of a 28-bp direct repeat element (DRE) found in the *tra*-2 3' UTR. The strongest *tra-2(gf)* mutation. *e2020,* is a 108-bp deletion that removes both copies of the DRE; *tra-2(e2020gf)* also feminizes the intestine (yolk, expression) and germ line (sperm and oocytes) of XO males. Northern blot analysis indicates that the steady-state level of *tra*-2 mRNAs expressed in *tra-2(gf)* mutants is not significantly elevated compared to wild-type animals. Two lines of evidence suggest that the tra-2(gf).

Mutants identify a binding site for a factor that mediates translational regulation of *tra*-2 mRNA through its 3' UTR. First, polysome gradient analysis indicates that tra-2 mRNAs isolated from *tra-2(gf)* mutants are associated with larger polysomes than wild-type *tra*-2 mRNAs, and second, the activity of a lacz reporter carrying the wild-type *tra-2* 3′ UTR is repressed in the hermaphrodite germ line, whereas the same reporter carrying the tra-2(gf) 3' UTR is not. A candidate for the DRE binding factor is encoded by the *laf-1* (for lethal and feminizing) gene, which decreases expression of the *lacZ-:tra-2* 3' UTR reporter in the germ line.

The molecular identity of *laf*-l is as yet unknown. The *tra-2(mx)* mutants feminize the germ line of′ XX hermaphrodites; however, unlike strong *tra*-2(gf) mutants, none of the *tra*-2(mx) mutants feminize XO males. The *tra-2(gf)* mutants Ire point mutations that result in nonconservative amino acid changes in a 22-amino acid region, named the NIX domain. The NIX domain is likely to identify a domain involved in a regulatory protein-protein interaction. Genetically, *her-1* and *fog*-2 are candidate repressors of *tra*-2 activity. However, XX *her-1* mutants develop as hermaphrodites, not females, and therefore do not affect the sperm-oocyte decision. In contrast, *fog*-2 mutations mimic the phenotype of *tra-2(mx) mutations:fog*-2 mutations transform XX animals into females, but have no effect on XO sex determination. Therefore, *fog*-2 may encode a candidate MX binding factor *fog*-2 has been cloned and shown to encode a novel protein.

Although both controls of *tra*-2 germ-line activity affect the onset of hermaphrodite spermatogenesis, it has not been established that translational control mediated by the 3' UTR normally plays a direct role in allowing the onset of hermaphrodite spermatogenesis. It could be argued that, because the *tra-2(gf)* mutants also affect XO males, translational control of tra-2 plays a more general role in maintaining TRA-2A protein at low levels. Otherwise, if TRA-2A expression is too high, its activity cannot be post-translationally regulated.

Chapter 5

Gene Expression

Homeobox genes encode proteins that have a homeodomain which is a 60-61 amino acid motif that was initially found in several *Drosophila melanogaster* homeotic genes. Since then, the homeobox has been found in all of the known homeotic genes in *Drosophila*, as well as in many other genes that regulate other aspects of development. In general, the homeodomain is the major conserved amino acid motif found in these proteins, although other motifs are found in some of these proteins. These include the Paired, *POU*, and LIM domains. Furthermore, zinc finger and leucine zipper motifs have been found on homeodomain proteins, although expression of these classes of homeodomain proteins has not yet been identified in the vertebrate brain. The homeodomain forms a helix-turn-helix structure that binds DNA in a sequence-specific manner.

The three-dimensional structure of the homeodomain bound to DNA has been solved for two homeodomains (*enlarged* from *Drosophila* and *MATa2* from yeast). Although the amino acid sequence of these peptides is quite divergent, their structures are nearly identical. These homeodomains have three a-helices and a flexible N-terminal arm. The structure of the second and the third helices are similar to the helix-turn-helix of bacterial repressor proteins. DNA sequence specificity is conferred by three residues in helix 3 and one to two residues in the N-terminal arm. The helix 3 side chains make base contacts to the major groove in DNA and the N-terminal arm side chain(s) interacts with the minor groove. Contacts with the deoxyribose-phosphate backbone are made by eight amino acid side chains; these residues are identical between *enlgailed* and *MATa2*, and are postulated to be responsible for proper positioning of the homeodomain on the DNA.

Homeodomain proteins are transcriptional regulators. They belong to a large family of helix-turn-helix proteins that are found in all organisms. In bacteria these proteins include the lambda, *trp*, and *lac* repressors. In yeast, homeodomain proteins *MATa1* and *MATa2* have pivotal roles in determining the dimorphic mating type phenotypes, whereas the PHO 2 protein is involved in phosphate metabolism. In all eukaryotes, these proteins take part in a wide range of functions; studied in greatest detail in *Drosophila melanogaster*. During embryogenesis of the fruit fly, homeodomain proteins play integral roles in setting up the embryonic axes, in establishing segmentation (gap, pair rule, and segment polarity genes), and in determining the identity of the segments (homeotic genes). Once the basic body plan has been determined, homeodomain proteins then have roles in controlling morphogenesis of body parts and in regulating cellular differentiation. All of the homeotic genes in *Drosophila* have a homeobox. The homeotic genes that specify the identity of the abdomen, thorax, and posterior part of the head are assembled in two clusters, called the *antennapedia-bithorax* cluster. The regulatory genes that specify the anterior part of the head (e.g., empty *spiracle* and *orthodenticle*) are not organized as it cluster. Null mutants of the *antennapedia-bithorax* genes lead to two general types of phenotypes: the deletion of a segment such as in null mutants of the labial and the deformed genes, and the transformation of a segment to a more posterior phenotype (a homeotic transformation) as exemplified by null mutants of the *sex combs reduced*, *antennapedia*, *ultrabithorax*, and abdominal A and B genes.

In the last several years, homologues of the genes in the *antennapedia bithorax* clusters have been identified in all metazoans. In most metazoans, these genes are organized as a single cluster. Evidence suggests that these genes subserve a central function in the development of all animals-the identity of cells along the anterior-posterior axis. Mutational analysis of these genes in nematodes, flies, and mice supports this hypothesis.

Vertebrates have four copies of the *antennapedia-bithorax* cluster. In mice, these genes are called the *Hox* genes, and the clusters are called *Hox* A, B, C, and D (the nomenclature used to be *Hox*-1, 2, 3, and 4). These genes are expressed along most of the longitudinal axis of vertebrate embryos into the posterior part of the head, analogous to their expression patterns in *Drosophila*. Genes at the 3' end of the cluster are expressed earliest and in the most anterior positions, whereas genes at the 5' end of the cluster are expressed later and in the most posterior position, There are 13 *Hox* paralogue groups that are believed to be evolutionarily related to a primordial *Hox* cluster. The mouse *Hox* clusters have between 9 and 12 *Hox* genes; thus, not all of the paralogue groups are represented in each cluster. In the central nervous system, these genes are expressed in the spinal cord and into the hindbrain, but do not reach the midbrain/ hindbrain junction.

Thus, as in *Drosophila*, a different set of genes is used to control regional specification of the anterior part of the central nervous system. This chapter describes the homeobox genes that are expressed during development of the mammalian brain. These include the *Hox* genes that are expressed in the hindbrain, as well as several other families of homeobox genes, most of which have homologues in Drosophila. These include the *Dbx, Dlx, Emx, En, Gbx, Gtx, LIM/homeobox, Msx, Nkx, Otx, Pax,* and *POU* genes.

During gastrulation the neural plate is formed along the dorsal side of the entire longitudinal axis of the embryo. The lateral edges of the plate, which give rise to the neural crest, fold toward the dorsal midline lining the ventricular surface of the neural tube form a pseudo-stratified epithelium called the ventricular zone. This cell layer contains the neural stem cells. When stem cells leave the cell cycle, they migrate away from the ventricular zone and begin to differentiate in the mantle zone of the neural tube. At specified positions along the longitudinal axis constrictions form in the wall of the neural tube. These constrictions are the morphological correlates, of hot boundaries that separate the principal structures of the central nervous system. i.e., the spinal cord, and the primary brain vesicles: the hindbrain (rhombencephalon), the midbrain (mesencephalon) and tile forebrain (prosencephalon).

Further morphogenetic events then modify each of the brain vesicles. This includes additional constrictions in the hindbrain and forebrain. In the hindbrain, these constrictions Subdivide the rhombencephalon into the large isthmo-cerebellar region rostrally and seven (or eight) transverse segments (rhombomeres) caudally. The midbrain and isthmus, which are not clearly separated morphologically at early stages, appear to be subdivided into several transverse regions that contribute to the mature midbrain and isthmic and cerebellar structures. Finally, the forebrain is also subdivided by transverse constrictions that may represent the boundaries between six forebrain segments called prosomeres. Within the entire neuraxis, there are four principal zones that are stacked ventrodorsally in parallel to the longitudinal axis.

The floor plate and roof plate are found at the ventral and dorsal midlines, respectively, and generally do not give rise to neurons. Between them are the major neurogenic plates, the basal (ventral) plate and the alar (dorsal) plate.

The longitudinal axis of the neural tube changes direction at three places. These bends occur at the junction of the spinal cord with the hindbrain (cervical flexure, ~90° flexion), in the middle of the hindbrain (pontine flexure, ~90° extension), and in the middle of the midbrain (cephalic flexure, ~180° flexion). It is critical to appreciate these changes in the direction of the longitudinal axis in order to understand neuromeric (segmentation) models of the brain. Available evidence suggests that differential rates of mitosis in basal and alar plates regulate, in part, the changes in curvature of the neural tube. Furthermore, the end of the rigid notochord near the junction of the midbrain with the forebrain may also contribute to the hairpin flexure at the midbrain-forebrain junction.

Additional morphogenetic changes further modify the brain. These are particularly complex in the prosencephalon. Upon closure of the rostral neuropore, the prosencephalon is a single vesicle called the primary prosencephalon.

The primary prosencephalon is subsequently subdivided by constrictions into transverse domains that may represent the forebrain neuromeres. From caudal to rostral the first three constitute the diencephalon, and the last three form the secondary prosencephalon. Additionally, a series of evaginations occur in the prosencephalon. These form the paired telencephalic vesicles, optic vesicles, and the preoptic and mammillary recesses, as well as the midline infundibulum and epiphysis. Regions in the telencephalon where there is rapid inward growth form prominent collections of mitotic cells that are called the lateral, medial, and caudal ganglionic eminences. As specification and morphogenesis take place, cellular differentiation and migration begin to generate the region-specific tissues of the forebrain. Mitotic zones are largely restricted to the periventricular layers of cells: the ventricular and subventricular zones. Cells that are programmed to leave the proliferative zones, migrate superficially and settle in the mantle layer where they undergo region-specific differentiation programs to produce particular cell types and intercellular connections.

The hindbrain in adult vertebrates consists of the medulla, pons, and isthmus structures that contain cranial nerves IV-II, as well its other important nuclei. Apart from the rostral isthmocerebellar region, the embryonic hindbrain (rhombencephalon) of all vertebrates displays seven neural segments called rhombomeres. Ail eighth rhornbomere has been postulated as well. Each segment contains a roof plate, a basal plate (where the motor neurons are located), an alar plate, and a roof plate. Hindbrain neural crest is derived from the dorsal edges of the alar plate. Neural crest begins to migrate away from the neural plate before it has completely closed and contributes to a wide range of tissues including the cranial ganglia, mesenchymal tissues in the branchial arches, the meninges, and the skull. During periods when regional specification is taking place, differential adhesive properties at the boundaries restrict the mixing of neuroepithelial cell clones between rhombomeres.

Furthermore, there are discontinuities of intercellular communication at the rhombomeric boundaries. Cells within a rhombomere are electrically coupled and allow diffusion of small dye molecules from one cell to another, whereas at rhombomeric boundaries there are cells that restrict the diffusion of small molecules. At this embryonic stage, it is hypothesized that specific patterns of gene expression specify the identity of each rhombomere. In particular, homeobox genes are expressed in the hindbrain in patterns that respect inter-rhombomeric boundaries.

To date, most of the work has focused on the *Hox* homeobox genes, but other classes of homeobox genes are also expressed in the embryonic hindbrain. The expression patterns of the *Hox* genes follow some general rules, but exceptions add complexity to this subject. The *Hox* genes are expressed in longitudinal domains extending through varying extents of the spinal cord. Genes from paralog groups 5-13 have their rostral expression boundaries in the spinal cord, whereas genes from paralogue groups 1-4 end in the hindbrain. The rostral expression boundaries of these *Hox* genes end at inter-rhombomeric boundaries. Genes at the 3' end of the *Hox* clusters are usually expressed in the most anterior positions. An exception to this rule is the *Hox-A2* gene; it is expressed up to the r1/r2 (r, rhombomere) boundary, rostral to the *Hox-A1* gene that is expressed up to the r2/r3 limit. Paralogue groups 3 and 4 from the *HoxA*, *B*, and *D* clusters have identical rhombomeric boundaries at r4/r5 and r6/r7, respectively, whereas paralogue groups 1 and 2 do not always share common boundaries. For instance, at E8 both *Hox-A1* and *Hox-B1* are expressed in a continuous domain up to the r3/r4 boundary; however, by E9.5 their patterns diverge. At this age, the caudal boundary of *Hox-B1* is modified so that its expression is restricted to r4, whereas the anterior boundary of *Hox/B1* moves rostrally to r2/r3. Therefore, *Hox-A1* expression extends one rhombomere rostral to *Hox-B1*. Likewise, *Hox-A2* is expressed up to r1/r2, whereas *Hox-B2* is expressed up to r2/r3. The latter gene becomes restricted to the r3-r5 segments later in development.

The expression patterns also show additional temporal and spatial complexities. At E8.0-E9.25, the level of expression of all the *Hox* genes is relatively uniform along the longitudinal axis. By E9.5, variations in the levels of expression are observed, as different rhombomeres have different levels of expression, and typically, the rostral-most domain has the highest level of expression. For example, *Hox-A3* and *Hox-B3* have the strongest expression in r5. on the other hand, the *Hox-A2* gene, which is expressed up to the r1/r2 boundary, has its highest levels of expression in r3 and r5. The expression patterns of the *Hox* genes are also spatially restricted along the dorsoventral axis; furthermore, these variations are dependent on the stage of development. For instance, up to E9.5 the *Hox-B* genes are uniformly expressed, but then many show a restriction to the dorsal spinal cord.

Krumlauf points out that this transition in expression occurs concomitantly with the appearance of specific cell types, and suggest that these genes may specify distinct cell types. On the other hand, *Hox-A2*, which is expressed dorsally in r2, cranial neural crest cells derived from r2 are probably not exposed to the *Hox-A2* protein. Crest cells from a given rhombomere migrate ventrally to populate the branchial arch mesenchyme that is at the same relative position along the longitudinal axis. Thus, crest cells from r1 and r2 fill branchial arch I (B 0, those from r4 fill B2, crest cells from r6 fill B3, and those from r7 and r8 fill B4. *Hox* genes are expressed in the crest cells emanating from r4, r6, r7, and r8 [there is some debate about whether rhombomeres 3 and 5 produce crest cells, although evidence suggests that they make some and that they have distinct migration pathways]. Thus, potentially, *Hox* gene expression in the hindbrain can contribute to the development of cranial ganglia and branchial arch-derived structures. Mutations of *Hox-A1* and *Hox-A3* bear out this hypothesis. No *Hox* genes are expressed in the crest cells that populate B1; thus other regulatory genes must be involved in the development of B1. Expression of the Dix-l and Dix-2 genes has been observed in neural crest derivatives in B1 (and in cranial ganglia) as well as in B2, B3, and probably B4. It is interesting that available data suggest that the Dix genes are not expressed in the hindbrain, implying that these genes are turned on only after the crest cells have left the neural tube.

Most of these genes are expressed in complex temporal and spatial patterns. Thus, it is likely that the regulatory pathways that control *Hox* gene expression are intricate. Several laboratories have initiated Studies into this problem by identifying regulatory elements using transgenic mouse assays. These are done by making a transgenic mouse that expresses a chimeric gene consisting of a fusion of a putative regulatory genomic fragment of a *Hox* gene to the *LacZ* coding region. Several transgenic mouse strains have been constructed that express *LacZ* in a pattern that is similar or identical to the endogenous *Hox* genes; these include strains expressing *LacZ* under the control of *Hox-A5, Hox-A7, Hox-B1, Hox-B2, Hox-B4, Hox-B6,* and *Hox-C8*. Systematic analysis of smaller parts of these regulatory domains allows the identification of the DNA sequences responsible for controlling various aspects of the expression patterns. Analysis of the *Hox-B2* gene has already led to the identification of DNA fragments that control expression in rhombomeres 3 and 5 and rhombomere 4.

Once putative enhancer and silencer elements are functionally identified, their DNA sequences can be studied. For instance, DNA foot printing analysis of the *Hox-B2* fragment that confers expression in rhombomeres

3 and 5 revealed a motif that is recognized by the *Krox20* protein. This was interesting because the *Krox20* protein is also expressed in rhombomeres 3 and 4. This finding was further studied using transgenic mice that express two transgenes. One transgene consisted of *LacZ* under the control of the *Hox-B2* fragment. The other transgene consisted of the *Krox20* coding sequence under the transcriptional control of the *Hox-B4* gene (*Hox-B4* is expressed in r7 and 1-8).

Thus, this transgene ectopically expressed *Krox20* in the caudal hindbrain (with it a boundary at r6/0). Therefore, in these mouse embryos, there were three domains of *Krox20* expression, the normal rhombomere 3 and 5 domains, as well as the ectopic domain in the caudal hindbrain. When the *Hox-B2-LacZ* expression was studied,)S-galactosidase activity was found in rhombomeres 3 and 5, and in the caudal hindbrain.

These studies strongly support the idea that the *Krox20* gene regulates the expression of *HoxB2* in rhombomeres 3 and 5. Mice homozygous for a loss-of-function allele of the *Krox20* gene have been constructed. Using molecular markers, it was concluded that rhombomeres 3 and 5 were missing in *Krox20* mutant embryos. Therefore, in this experiment it was not possible to test whether *Hox-B2* expression was affected by the lack of *Krox20*.

Thus, while this approach is making inroads into understanding how other transcriptional regulators can control expression of *Hox* genes, they do not address how spatial information can regulate these genes. For instance, what is the substance(s) that is produced by Spemann's organizer that travels along the plane of the ectoderm and induces neural tissue and transverse bands of homeobox and *Krox20* gene expression? How is this signal(s) transduced to alter gene expression? There are many other examples of inductive and morphogenetic processes in neural development for which there are few data regarding their molecular mechanisms. However, retinoic acid is a candidate for having a role in several of these processes. Furthermore, retinoic acid can effect the expression of the *Hox* genes. Retinoic acid and/or chemical relatives of this molecule have been implicated in a wide variety of embryological processes including limb development, gastrulation, and olfactory bulb induction. The initial evidence that retinoic acid can regulate homeobox gene expression came from experiments with embryonal carcinoma cells. Treatment of these cells with retinoic acid leads to stereotypic responses in the expression of the *Hox* genes.

Genes at the 3' end of the clusters are Induced first and reach maximal levels of expression within about 24 hr. Genes at the 5' end of the clusters are either induced late or weakly; expression of some of the 5' genes also show inhibition by the retinoic acid. For instance, genes in the middle of the clusters reach maximal levels of expression around 60-80 hr after exposure to retinoic acid, and some genes at the 5' end of the clusters do not reach maximal expression even after 180 hr. These cell culture experiments have been extended by studying the expression of *Hox* genes. In retinoic acid-treated embryos. Retinoic acid is it potent teratogen, its effects are complex because they are dependent it the dose and stage of the treated embryo. Retinolic acid exposure of all vertebrate species resu!ts in morphological abnormalities of the hindbrain. These abnormalities appear to be due in part to a posteriorization of the hindbrain. Furthermore, the expression of the *Hox* genes in retinoic acid-treated embryos is abnormal; the rostral boundaries of the genes are more anterior than normal. Analysis of the effect of retinoic acid on homeobox gene expression has been aided by the use of several transgenic mouse strains that express the *LacZ* reporter gene under the control of different regulatory sequences: *Hox-B1*, *Hox-B2*, *Krox20*, and *Hox-A7*. For instance, retinoic acid treatment of mouse embryos harbouring the *Hox-B1-LacZ* gene at E7.5 causes an anterior shift in the expression domain extending through the entire rostral hindbrain at E8.5 (the expression in the hindbrain at E8.5 of untreated embryos is restricted to rhombomere 4). Examination of the retinoic acid treated embryos at E9.5 revealed a refinement of the pattern such that LacZ was found in rhombomeres 2 and 4; again untreated embryos at this age expressed *LacZ* only in rhombomere 4. This showed that retinoic acid caused a duplication of the rhombomere 4 *Hox-B1* domain in rhombomere 2, implying that the ectopic expression of *Hox-B1* and perhaps other genes leads to the respecification of rhombomere 2 as rhombomere 4. This was substantiated by the analysis of expression patterns of the *Hox-B2* transgenes. Furthermore, examination of the expression pattern of the *Krox20* transgene suggested that rhombomere 3 had been respecified as rhombomere 5.

Analysis of the patterning of motor neurons and axon pathways in these rhombomeres also supports these conclusions. These studies clearly demonstrate that retinoic acid changes the phenotype of rhombomeres 2 and 3 into structures that have the same gene expression patterns, motor neuron organization, and axon trajectories as those seen in rhombomeres 4 and 5. They suggest, but do not prove, that the changes in *Hox* gene expression have a role in respecifying the hindbrain neuroepithelium. Direct evidence for the role of the *Hox* genes in hindbrain development has come from a mutational analysis of these genes, which is described later. At this juncture, it is important to mention that there are many other homeobox genes that are expressed in the embryonic hindbrain.

Most of the known genes appear to be expressed continuously along the longitudinal axis, including several *POU* and *Pax* genes. There are additional homeobox genes that are expressed in spatially restricted domains, including *Gbx-2*, roughly in rhombomeres 6-7 and 1-2 and EA-1, in the rostral rhombencephalon.

Loss of function mutations of four marine *Hox* genes, produced using gene targeting methods, have been published. Two of these mutations affect genes that are expressed in the hindbrain. In each case, the regions of the embryo affected by these homozygous mutations (heterozygotes appear normal) correspond to the anterior part of the gene expression domain; for the *Hox-A3* mutation, structures anterior to its expression domain were also disrupted. The *Hox-A1* and *Hox-A3* mutations did not produce homeotic transformations; they generated malformations and perhaps deletions of certain Structures. Mutation of *Hox-A1* has been studied by two laboratories. Their major findings were similar while there were some minor differences in the phenotypes. This may be due to differences in the deletions that they made in the *Hox-A1* gene, which produces two transcripts, one of which appears not to encode the homeodomain. The n-tutation generated by Lufkin et al. should eliminate both transcripts, whereas the Mutation made by Chisaka et al. eliminates only the transcript encoding the homeodomain.

In both mutations, heterozygotes were normal whereas homozygotes died at birth and had a range of defects in the part of the hindbrain derived from rhombomeres 4-7 (r4-0), as well as defects in the derivatives of the neurogenic neural crest, the inner and middle ear, and the occipital bones of the skull. The defects included absence of the motor nucleus of the 7th nerve, part of the acoustic ganglia of the 8th nerve, the roots of the 9th and 10th nerves, and the superior olivary complex. Chisaka et al. observed that the morphology of the rhombomeres was disrupted, and Lufkin et al. found that the mutation caused delayed closure of the rostral neural tube.

Both laboratories have reported more detailed analyses of the anatomy and gene expression patterns of mutant embryos. Again, there were similarities and differences in the phenotype of the mutants. Analysis of the gene expression patterns of certain *Hox* genes, *Int*-2 (FGF-3) and the *Krox20* genes, at E8.5-E9.5 in the rhombencephalon of *Hox-A1* Mutant embryos clearly shows abnormalities in rhombomeres 4 and 5. Both groups found that rhombomere 5 was largely or completely deleted.

Carpenter *et al.* interpreted their results as a total deletion of rhombomere 5 and a decrease in the size and a reduced level of *Hox-B1* expression in rhombomere 4. Dolle et al. interpreted their findings as a deletion of most of rhombomeres 4 and 5, although vestiges of each could be detected. For instance, they found that expression of *Hox-B1*, which is limited to rhombomere 4 at E9.5, is reduced to a small cluster of cells at the dorsal r3/r4 border. Likewise, the r5 expression domain of *Krox20* is reduced to a thin dorsal cluster of cells at the rostral boundary of r6. Associated with this size reduction of r4 and r5 are morphological alterations including the absence of two rhombomeres, the rostral displacement of the otic vesicle, disruption of the facial acoustic ganglion (which is formed in part by neural crest cells derived from r4), and the lack of the neural crest-free area flanking r-5. Dolle et al. suggest that r4 and r5 are deleted in the *Hox-A*1 mutant, and propose that the absence of *Hox-A1* causes either apoptosis or alterations in growth control of cells in r4 and r5. Because there are thin strips of cells expressing r4 and r5 markers between r3 and r6, they point out that specification of these rhombomeres has not been entirely disrupted. It is interesting that mutations in the *Pit-1* homeobox gene also lead to decreases in the number of cells that are believed to be specified by expression of this gene. Carpenter et al. found that the wiring of the mutant hindbrains was abnormal. For instance, they discovered that in mutant embryos, cranial nerves VII and VIII receive efferent axons from rhombomeres 6-8; these nerves normally contain efferents only from cell bodies in rhombomeres 4 and 5. Therefore, while the main embryonic field disrupted by the *Hox-A1* mutation is rhombomeres 4 and 5, more posterior parts of the hindbrain are also affected indirectly by this genetic defect. Mutation of the *Hox-A3* gene produced a surprising phenotype because, although it is expressed in the embryonic hindbrain, abnormalities were restricted to cranial neural crest derivatives. Furthermore, the cranial neural crest defects did not disrupt the peripheral nervous system, as had the *Hox-A1* mutation. Instead, the mutation affected mesenchymal and endocrine development, including disruption of the thymus, parathyroid, thyroid, heart, and great vessels. Mutations of the *Hox-B4* gene, which is expressed along the spinal cord into the hindbrain in rhombomere 7 and in the nodose ganglia, does not have a readily detectable phenotype in any neural structure. Although the nervous system is not apparently affected, this mutation does cause a morphological change in the cervical vertebrae that is consistent with a homeotic transformation.

Homeobox genes in the midbrain, isthmus, and cerebellum

In the adult, the midbrain, isthmus, and cerebellum consist of alar structures, such as the superior (tectum) and inferior colliculi, the parabigeminal and parabrachial nuclei, and the cerebellum, as well as basal structures, such as the red nucleus, substantia nigra, and the motor nuclei for cranial nerves III and IV. In the embryo, this brain region is intercalated between the rhombomeres and prosomeres. At early stages, it does not have distinct morpho-logical features that clearly subdivide it into distinct segments. This has led to controversy and confusion regarding the organization of this region. However, cytoarchitectural patterns, fate maps, and gene expression patterns distinguish several parts that are disposed serially along the longitudinal axis.

In order to clarify the relationship of this model to standard anatomical nomenclature, it is useful to point out how the pons can be understood within the context of the model in Fig. 4. The mature pons, or metencephalon, is a region that is characterized by ventrally prominent structures, including the middle cerebellar peduncle and pontine nuclei. It also includes the trigeminal main motor and sensory nuclei which are derived from rhombomeres 1-3. Classic definitions of the pons generally include the cerebellum as its dorsal structure. However, there are inconsistencies with this definition. For instance, the cerebellum largely originates rostral to r1, from the *En-2* positive domain. Also, some definitions of the human pons also include the nuclei of the VI and VII cranial nerves, though these originate in r4-r6. This false impression is due to the topological deformation of the massive pontine region found in the adult human brain. Therefore, there is nonconcordance of the classical concept of the pons (metencephalon) with the modern neuromeric schemata. It is our opinion that the neuromeric conception agrees more completely with experimental observations and suggest that this terminology should be used.

There are two homeobox genes whose expression is largely restricted to this brain region: *En-1* and *En-2*. Many other homeobox genes are also expressed in this region, but the midbrain, isthmus, and/or cerebellum do not make up their principal domains. The *En-1* and *En-2* genes (homologues of the *Drosophila engrailed* gene) have been studied in the greatest detail. In *Drosophila*, the engrailed gene has a homologue called invected; these genes are closely linked. *En-1* and *En-2* are not linked; *En-1* is on the central part of mouse chromosome I and *En-2* is on the proximal part of mouse chromosome 5. Like the engrailed and invected genes, *En-1* and *En-2* are expressed with similar patterns during embryogenesis. The expression of these genes has been studied in zebrafish, frog, chicken, and mouse. These comparative studies reveal that the expression patterns have been highly conserved between these vertebrate species.

En-1 and *En-2* expression begins at about E8.0 in a transverse band of neuroepithelium th!Lt encircles the neural tube at the mesencephalon-rhombencephalon junction. An interesting aspect of this is that expression of the growth /differentiation factor *Wnt-1* may begin a few hours before *En-1/2* in this same domain (*Wnt-1* has additional domains of expression where the engrailed homologs are not expressed). There is accumulating evidence that as in *Drosophila*, the *Wnt-1* gene may have a role in the regulation of En expression. This point will be amplified later. *En-1* and *En-2* are expressed in a bidirectional gradient, with maximal expression at the isthmus, and decreasing expression rostrally into the mesencephalon (perhaps extending at least to the m/pI boundary) and caudally into the cerebellar part of the rhombencephalon.

From E8 to E 11, the *En-1* and *En-2* patterns are largely indistinguishable, but at E12 their expression diverges. En-1 expression becomes weaker than *En-2*, and En-l forms new expression zones both within and outside the nervous system it is expressed in two longitudinal stripes along the basal plate of the hindbrain and spinal cord, these cells probably represent the primordia of motor nuclei inasmuch as the trigerninal motor nucleus clearly expresses *En-2* at E17.5. Furthermore, there are differences in their expression in r1 and in the primordia of the cerebellum. Their expression late in gestation and Into adulthood further diverges as exemplified by the fact that only *En-2* is expressed throughout development of the cerebellar and isthmic regions. Both *En-1* and *En-2* are also expressed in the periaqueductal grey internal to the inferior colliculus and in the substantia nigra (m2) in the adult mouse.

Information regarding the function of the vertebrate *En* genes has come from the analysis of the *En-2* mutation and experimental embryological manipulations. The gene targeting method was used to create a mouse strain with a null mutation in the *En-2* gene. Mice that do not express *En-2* have a subtle alteration in the structure of the adult cerebellum, characterized by abnormal foliation and no clear neurobehavioural phenotype. Thus, it is possible that *En-2* and *En-1* (and perhaps other genes) share redundant functions; the cerebellar phenotype may have arisen because *En-1* is not expressed in the laterocaudal part of this structure late in gestation. The function of the *En* genes has also been studied using an indirect approach: through observing the effects of ectopically transplanting *En* expressing neural tissues. There have been several general approaches utilized in these experiments. Nakamura and colleagues transplanted chick embryo tecta into the caudal diencephalon. They found that the ectopic tecta continued to express *En*, but did so in a gradient opposite to the normal mesencephalic caudal to rostral gradient. They found that the closer the mesencephalic graft was to the mesencephalic/diencephalic junction, the lower the expression of *En*. These results provided evidence for a repressive influence on *En* expression from the mes/dien boundary; presently there are no candidate molecules for this repressor. Furthermore, the observation that *En-2* expression normally continues into the diencephalon along the basal plate, but not the alar plate, suggests that the postulated repressor only operates in the alar plate. It was also found that the cytoarchitecture of the ectopic tecta resembled the normal tecturn.

Very little is know about the extracellular and intracellular molecules that regulate *En* gene expression. A potential inroad to this area is based on information from *Drosophila,* where the wingless protein plays an important role in the induction and maintenance of engrailed expression. Consistent with this, there is indirect evidence that one of the vertebrate wingless homologs, *Wnt-1,* may regulate *En* gene expression. This hypothesis is based on the following reasonings: (1) *Wnt-1* is expressed just before *En-1 En-2* at the mecencephalon/ rhombencephalon (mes/rhom) junction; (2) transplantation of different transverse slices from the mes/rhom junction into the diencephalon induces the adjacent diencephalic tissue to express *En;* the slices that induce the most En conic from the region where *Wnt-1* is expressed the highest, (3) *Wnt-1* mutations (two via gene replacement and one spontaneous mutation called swaying) result in phenotypes that range from deletion of the entire *En1/En-2* expression domain to loss of only the middle of the *En* domain. Experimental embryological manipulations invol-ving ectopically transplanted pieces of isthmus into the forebrain identified the isthmus as source of an agent, perhaps *Wnt-1,* that can induce *En* expression. Conversely, when mesencephalic tissue was transplanted into the diencephalon, the level of *En* expression was reduced in the part of the ectopic tissue that was closest to the mesencephalic/diencephalic boundary: Martinez et al. also found that transplantation of such prospective isthmocerebellar tissue into the rostral diencephalon resulted in maintenance of *En* expression in the ectopic tissue as well as induction of *En* expression in the adjacent diencephalic cells. The transplant developed into a cerebellar-like structure. However, when *En*-expressing prospective isthmo-cerebellar tissue was transplanted into the rostral mesencephalon, near the mesencephalic/diencephalic boundary, the level of *En* was reduced and the tissue developed into a mesencephalic phenotype instead of that of a cerebellum, suggesting that the mesencephalic/diencephalic boundary is a source for a repressor of *En* expression. Of note, induction of *En* expression in the diencephalon does not extend across interprosomeric boundaries, as observed for the zona limitans and the pretecto-thalamic boundary, and possibly respects the alar/basal boundary.

This suggests that boundary cell populations may be nonpermissive for the propagation of the signal transduction cascade necessary for diffusion of the *En*-inducing agent. Curiously, only pretectal and dorsal thalamic prospective portions of the diencephalon are susceptible to *En* induction. The rostral-most forebrain, encompassing ventral thalamus and telencephalon, supports the differentiation in cerebellum of the *En*-2 positive grafts, but does not express *En*-2 nor change its differentiation pattern. Thus, these experiments provide evidence for the existence and the location of factors within the embryonic brain that induce or repress *En* expression. They also provide indirect, yet suggestive evidence that correlates the ectopic expression of *En* with a change in phenotype.

Marin and Puelles have shown that inversion of a neuro epithelial ring containing the mesencephalon, isthmus, and cerebellar primordia controls correct histogenetic differentiation of all midbrain and isthmic centers in adjacent neuroepithelia that are susceptible to *En*-2 induction. The histogenic patterns are polarized according to the relative position of the isthmocerebellar ring. It is postulated that the polarizing activity underlies the rostrocaudal gradient of maturation in the tecturn and other discrete differentiation differences normally found in the midbrain and isthmus.

Therefore, this line of investigation provides strong evidence for the presence of an organizer-like structure at the midbrain-hindbrain boundary that secretes morphogenetic signals which pattern histodifferentiation in adjacent neuroepithelial domains. It also suggests that patterning in other CNS regions may be orchestrated by additional morphogenetic organizers. These experiments also implicate the En homeobox gene in this process and Suggest that ectopic expression of this gene in discrete locations within the embryonic brain could be used to test this hypothesis.

While there is some evidence for the existence and possible identity of diffusable factors that induce *En* expression in permissive cells, information regarding surface receptors, signal transducing machinery, and transcription factors that induce or repress *En* expression is lacking. Inroads to this are beginning to be made by identifying the regulatory elements that control the temporal and regional specificity of expression of the *En-2* gene. Two enhance r-containing regions upstream of the *En*-2 transcribed sequences have been found to control tissue-specific expression. A 1.5-kb fragment contains an enhancer that controls expression in the midbrain/hindbrain region, and a 1.0-kb fragment regulates expression in the mandibular myoblasts.

The forebrain consists of the diencephalon, and secondary prosencephalon. As postulated recently in our forebrain model, the secondary prosencephalon includes the whole hypothalamus, the eye stalks and vesicles, the preoptic region and commissural plate, the telencephalic peduncle, and the telencephalic vesicles. The telencephalon is the largest and most complex part of the mammalian brain: it contains the cerebral cortex, basal ganglia, olfactory bulbs, and other structures. Several classes of homeobox genes are expressed in the embryonic forebrain, or prosencephalon. Some of these are only expressed in restricted fore-brain domains, whereas others are expressed throughout most of the nervous system. To date, mutations have been identified in only two of these genes: thus most of this section will describe the expression patterns of these genes. Like the expression of the *Hox* genes in the hindbrain.

The expression patterns of many of these genes show sharp transverse and longitudinal boundaries that are reminiscent of the *Hox* expression at rhombomeric boundaries. Using these and other data, a neuromeric model of forebrain organization that allows detailed morphological mapping of gene expression patterns, eschewing the difficulties posed by conventional schemata has been proposed. This schema proposes that the forebrain is subdivided into transverse domains called prosomeres (p).

POU genes (*Pit/Oct/Unc*) encode proteins with two functional domains: the *POU* domain and a homeodomain. These motifs are colocalized within a stretch of 150-160 amino acids: the aminoterminal 75-82 amino acids are the *POU*-specific domain: this motif is followed at a conserved linker region that separates the *POU*-specific domain from the 60 amino acid homeodomain. Based on the amino acid sequence of the linker region and the amino-terminal part of the homeodomain. The mammalian *POU* proteins are categorized into six groups: I (Pit-1): 11 (*Oct*-I and *Oct*-2): III [*Brn*-1, *Brn*-2. *Brn*-4. and *Oct*-6 (Tst-I and *SCIP*)]: IV (*Brn*-3.0. *Brn*-3. 1. and *Brn*-3.2): V (*Oct*-3/4): and VI (*Brn*-5). Presently, there are 13 known mammalian *POU* genes, 11 of which are expressed in the developing and adult central nervous system as well as many other parts of the mouse. The best characterized mammalian *POU* gene is *Pit-1*. It is expressed in the embryonic pituitary, and will be discussed subsequently.

Most of the cther *POU* genes are widely expressed during embryogenesis in the proliferative zone of the CNS. Due to the limited published information on the embryonic expression patterns of most of these genes, it is difficult to make firm conclusions about specific boundaries, Therefore, our interpretations are tentative. *Brn*-1, *Brn*-2, *Brn*-3, and *Tst*-1 are expressed in all of the visible parts of the E13 rat CNS. At E16, *Brn*-1 is expressed in most of the CNS, although the anterior entopeduncular, preoptic, chiasmatic, and tuberal areas are negative. *Brn*-2 expression at E16 is also extensive, although there is a gap caudal to r8 that may include the two most rostral myelomeres. *Brn*-3.0 and *Brn*-3.2 have a much more restricted expression pattern at E16. Data on *Tst-1* (*SCIP* or *Oct*-6) show that it is expressed in the embryonic telencephalon at E16, as well as in a caudal-rostra] gradient in the alar mesencephalon. *Brn-4* expression maps in the basal plate tip to p4/p5 (tuberomanimillary boundary). Its rostral alar limit is difficult to discern, but may stop at either pl/p2 or p2/p3. Postnatally the various *POU* genes continue to show restricted expression in specific neuronal populations. Thus, *Brn*-1 and *Brn*-2, which have the most widespread expression patterns, appear in the inferior and superior olives, motor and interpeduricular nuclei as well as the cerebellum and isthmic alar formations of the brainstern. Expression is also found in tile alar and basal midbrain (m1, m2), medial habenula (p2), and ventral thalamus (reticular nucleus and zona incerta (p3).

Both genes are expressed throughout the cerebral cortex and the olfactory bulb (p4-p6), in various basal telencephalic populations like the nucleus on the lactal olfactory tract, the islands of Calleja, medial septum, diagonal band, substantia innominata, and the bed nucleus of the stria terminals as well as in the lateral mammillary and tuberomammallary nuclei, other posterior hypothalamic formations, the supraoptic/ paraventricular complex (p4), suprachiasmatic nucleus, and medial preoptic area (p6). Their expression is excluded from the amygdala, striatum, and main centers of the pallidum. In contrast to the widespread expression of *Brn*-l and *Brn*-2, *Brn*-3.0 and *Brn*-3.2 expression is more restricted. These genes are also expressed in both the embryo and the adult brain, and appear to be expressed primarily in postmitotic cells. These *POU* genes have the greatest homology to the C. elegans *Unc*-86 gene. *Unc*-86 has a role in differentiation of hermaphrodite-specific motor neurons and mechanosensory neurons. *Brn*-3.0 RNA is found in sensory ganglia, inferior olive, and the ambiguous motor nucleus in the caudal brainstem. It is also expressed in a large transverse domain in the isthmus and both inferior and superior colliculi, as well as in a set of cells in the medial and lateral habenular nuclei (p2) in the posterior hypothalamus pituitary and the retina. Gerrero et al. present evidence that *Brn*-3.0 regulates the expression of the proopiomelanocortin gene. *Brn*-3.2 shares similar expression domains with *Brn*-3.0: its RNA is found in the spinal cord, spinal and trigeminal ganglia, inferior olive, superior colliculus, inter-peduricular nucleus, and the retina.

Brn-4 expression becomes restricted in advanced fetuses to the subcommissural organ (p1), the medial habenula (p2). the striatal subependymal layer and supraoptic/paraventricular complex (p5). nucleus accumbens (p6). and possibly the mammillary complex (p4). Other *POU* genes show similarly discrete expression in adult animals, including *Oct*-6 (*SCIP*. Tst-l) expression in specific cell types in the cerebral cortex. *Oct*-1 expression in cerebellar granule cells: and *Oct*-2 expression in the suprachiasmatic and medial mammillary nuclei. Furthermore, *SCIP* (*Oct*-6, *Tst*-1) is expressed in the oligodendrocyte and Schwarm cell lineages. It is induced by cyclic AMP and during periods of rapid cell division.

The *Pax* genes have a paired box which is a 128 amino acid motif that is homologous to the DNA-binding domain of the *Drosophila pairedprotein*. Presently. there are 10 known *Pax* genes in mammals (8 in mice); most of these genes encode proteins that also contain homeodomains. Unlike the *Hox* genes, the *Pax* genes are not organized into multigene clusters. Several of these genes are known to be expressed. In the forebrain, although these also have extensive expression domains in more caudal regions of the central nervous system and in other parts of the embryo.

The expression of the *Pax* genes can be used to divide them roughly into three groups. *Pax*-3 genes begin to be expressed at -E8-E8.5 in neuroepithelial cells. Their patterns of expression are very similar as they are found in alar territories that extend longitudinally throughout the spinal cord, hindbrain, and midbrain into the caudal diencephalon. Their rostral limits are at the p1/p2 boundary; although *Pax*-3 may extend into the epithalamus, a region that we include as the most dorsal alar domain in p2.

Although both genes are expressed in the dorsal neural tube, only *Pax*-3 is expressed in the roof plate (the dorsal midline). The expression of *Pax*-2, *Pax*-5, and *Pax*-8 begins slightly later in development than *Pax*-3 and *Pax*-7(at ~E9.5—E10). at a time that coincides with early neurogenesis. Note that these genes have no homeobox. Genes of this second class are expressed in neuroepithelial cells and in neurons. All show expression domains that cover specific alar and basal territories. *Pax*-2, *Pax*-5, and *Pax*-8 have rather similar patterns, characterized by a rostral domain that is centered in the isthmic region, and a more extensive caudal domain that extends along the entire spinal cord into the hindbrain up to the pontine region (at E 11. 5 *Pax*-2 and *Pax*-8 seem to stop at the r2/r3 boundary, whereas *Pax*-5 apparently stops at the r3/r4 boundary). *Pax*-2 is also expressed in the ventral retina and ventral eye stalk.

These three genes also exhibit distinct expression patterns within the isthmic domain in patterns that are similar to the En genes. *Pax*-2 and *Pax*-8 transcripts predominate in the dorsolateral alar region of the rostral cerebellar and isthmic "segments" and expression ends abruptly at the Ist/m2 boundary. On the other hand. *Pax*-5. whose expression in the caudal hindbrain and spinal cord is weaker and restricted to fewer neurons has a more intense and extensive rostral domain including again the cerebellum and the whole isthmus and the caudal-most part of m2. Moreover, *Pax*-5 extends into the ventrobasal isthmic regions. Its domain within the midbrain basal plate may be more extensive rostralward, perhaps reaching p-1. Functional analysis of these genes is quite limited, although injection of antibodies against the zebrafish *Pax*-2 homologue into zebrafish embryos has been shown to disrupt expression of the zebrafish homologues of *Wnt*-1 and *En*-2 as well as development of the isthmic region. The *Pax*-6 gene makes Lip the third *Pax* group. Its transcripts are the most apparent in a rostral domain, within the alar region of the forebrain. This domain has a sharp caudal limit that coincides with the m1/p1 boundary. It extends continuously throughout alar p1 and p2. stopping just short of the boundary that separates p2/p3 (the zona limitans).

Expression extends more rostrally within alar p3 and p4 (excluding the caudal ganglionic eminence) and then continues in a thin longitudinal band through p5 and p6 (within the anterior hypothalamic and posterior preoptic domains) into the optic stalk and eye vesicle. Expression within p5 and p6 is also found in the cerebral cortex, olfactory bulb, septum, and diagonal band domains. Note that *Pax*-6 is also expressed in a separate caudal domain within the basal plate of the Cb, r1-r8, and the spinal cord. At later developmental stages, *Pax*-6 transcripts diminish in the dorsal thalamus (alar p2) and in p4-p6, but remain high in the ventral thalamus (alar p3). Finally, a novel domain appears between E15.5 and E18.5 in the rhombic lip and its derivatives (external granule layer of the cerebellum and pontine and olivary nuclei). The expression patterns of the *Pax* genes have been conserved across vertebrate evolution. For instance, the expression of *Pax*-6, which is the *Pax*-6 homologue in zebrafish, is very similar to the mouse *Pax*-6 gene. Expression is found in the basal plate of the spinal cord and hindbrain, as well as in the alar plate of the forebrain. Within the forebrain, the published data indicate that the interruption of expression within the p2/p3 interprosomeric boundary arises secondarily. Initially, there is a continuous alar domain from the m1/p1 boundary to the optic stalk. There is no expression in the telencephalon, at least at the published embryonic stages. *Pax*zf-b] expression is similar to *Pax*-2. Both have separate domains in the eye vesicle and stalk, in the isthmornesencephalic boundary, and in the caudal brainstern and spinal cord.

The functions of several *Pax* genes have been studied using naturally occuring mutations. Mutations have been identified for three *Pax* genes in mice and in two *Pax* genes in humans. *Pax*-1 mutations (called undidated in mouse) affect the vertebral column, and will not be discussed because this gene is not expressed in the CNS. On the other hand, homozygous mutations in *Pax*-3 cause severe malformations of the central nervous system that have been primarily described as dysraphias, or neural tube closure defects, and abnormalities of tissues that contain neural crest. Heterozygotes have white spotting of their abdomen, tail, and feet ("splotching"), presumably due to an abnormality of melanocytes that are neural crest derivatives. Mutations in the human *Pax*-3 gene result in Waardenberg's syndrome, characterized by pigmentation disturbances (e.g., a white forelock) and sensorineural deafness. At this point, deficits in hearing have not been identified in the mouse *Pax*-3 mutants.

Mutations of the *Pax*-6 gene have also been identified in mice and humans. In mice, heterozygotes for the mutant allele are viable and have small eyes; this is the basis for the name of the mutation: small eye or sey. Homozygotes are inviable, probably because they develop no nasal cavities. They also do not develop eyes. The mechanism for these defects is uncertain, but may involve abnormal induction of the nasal and lens placodes.

A defect in the migration of midbrain neural crest has been detected in *Pax*-6 mutant rats, these cells normally migrate throughout the face, but in the mutant they did not migrate to the nasal rudiment, suggesting that this gene plays a role in nasal placode induction. Humans lack an iris, when they are heterozygotes for a mutation of *Pax*-6, in a syndrome called aniridia. As described earlier, *Pax*-6 is widely expressed in the embryonic CNS, yet mutations of this gene apparently only affect development of a small subset of these tissues, although a detailed analysis of the brains of these mutants remains to be published.

The *Gbx-2* gene (gastrulation-brain-homeobox, also known as MMox-a) does not have a known Drosophila homologue. It is related to another murine gene called *Gbx-1*. *Gbx-1* and *Gbx-2* are located on mouse chromosomes 5 and 1, respectively. *Gbx-2* is expressed during neurulation in the neural ectoderm and underlying mesoderm with a rostral boundary in the region of the midbrain-hindbrain junction (M. Frohman and G. Martin. unpublished results). By E12.5 it is expressed in two longitudinal stripes in the spinal cord up to the r6/r5 boundary, and has additional separate domains in rhombomeres r2 and r1, in the dorsal thalamus (alar p2). and in a superficial ceH population of the medial ganglionic eminence (MGE). Expression of *Gbx-1* has not been detected in the brain at E12.5.

The *Dbx* gene does not have a known Drosophila homologue: it is expressed longitudinally in a thin strip of subventricular and mantle cells along a line that may correlate with the alar/basal boundary. This expression domain extends across the spinal cord, hindbrain, midbrain, and may extend to the p2/p3 boundary in the diencephalon, although the published illustrations do not clearly resolve this latter point. This domain, or perhaps a separate domain, is also present at the alar/basal boundary of p3 and ends rostrally in a wider zone corresponding to the mammillary area (basal p4). Additionally. *Dbx* is expressed less intensely in the alar plate of the m1, m2, pl and p2 segments, being sharply limited at the m2/1st and p2/p3 boundaries. A smaller separate domain maps in the telencephalon to mantle cells superficial to the lateral ganglionic eminence (LGE), a region that may become the claustrum.

The *Gtx* gene encodes a highly diverged homeodomain with no known Drosophila homologue, which is also unusual because of its DNA-binding specificity. Unlike most homeodomain proteins. *Gtx* does not efficiently bind to a variety of standard oligonucleotide sequences such as (TAA)5. Rather, it binds to a consensus binding site named MEF-2, which is found in a variety of enhancers, such as in the muscle creatine kinase gene. *Gtx* shares binding specificity with a family of transcriptional regulators that share a structural motif called the MADS box. *Gtx* is expressed at h Igher levels in the adult than in the embryonic CNS. At E18. expression is seen in the spinal cord, hindbrain, midbrain, and hypothalamic area. In the adult brain. *Gtx* is expressed in the white matter in astrocytes and oligodendrocytes.

The LIM/homeodomain genes encode proteins with two structural motifs: the LIM and homeodomains. The LIM motif corresponds to a conserved polypeptide domain found in the *Lin-11, Is/-1,* and *Mec-3* regulatory proteins that control cell fate determination in C. elegans (*Lin-11* and *Mec-3* genes) and bind the insulin I enhancer region (*1sl-1*gene). The LIM motif encodes metal-binding cysteine and histidine-rich domains of unknown function. LIM domains are also found in a *Drosophila* homeobox gene called apterous. Its well as in the rhombotin protein that Is expressed in various CNS locations. Rhombotin is implicated as a T-cell oncogene.

To date, two mammalian LIM/homeodomain genes have been reported to be expressed in the brain. *LH-2* is expressed in embryonic and adult forebrain and midbrain regions, its well its in developing lymphocytes. The lsl-l gene is expressed in subsets of neurons and endocrine cells in the adult rat. It is found in peripheral organs, including the sensory ganglia, its well its in the CNS, where it is expressed in subsets of spinal cord and brainstern motor nuclei, and probably in some catecholaminergic neurons. Expression in the forebrain is remarkable because it is found in cells that are derivatives of neurciepithelial domains that express *Dlx*-1 and *Dlx*-2. Expression is found in derivatives of the ventral thalamus (alar p3, the reticular nucleus and zona incerta), hypothalamic anlage (alar p4-p6-, paraventricular/supriloptic, suprachiasmatic, and arcuate nuclei), and basal telencephalon. Of note, no expression is found in the optoeminential domain where the *Dlx* genes are not expressed. Finally, a *Xenopus* LIM/homeodomain gene hits also been reported to be expressed in embryonic and adult CNS structures. The LIM/homeodomain proteins are interesting because, like many of the *POU* proteins and some other homeodonlain genes, they are expressed in differentiated cells that have left the mitotic cycle (postmitotic cells). Therefore, it is likely that they are downstream targets of homeodomitin proteins that are expressed at earlier stages of differentiation.

There are two genes that are homologous of the *Drosophilla* orthodenticle gene: *Otx*-l and *Otx*-2. The orthodellticle gene regulates development of specific structures in the head of *Drosophila*. The homeodomains of the *Otx* class are distinctive because they are similar to the Drosophila bicoid protein. The bicoid protein is required for the initial steps of A-P patterning of the *Drosophila* embryo and regulates expression of the orthodenticle gene, hence there is great interest in whether the otxgenes have similar functions to the bicold and orthodenticle genes. *Otx*-2 is expressed in embryonal stem cells and is found at E5.5 in the entire embryonic ectoderm. However, during gastrulation, expression disappears posteriorly and becomes restricted to the anterior region of the ectoderm: the primordia of the mid and forebrain. By E8.5, Otx-I expression also begins in it large interior- neuroectodermal domain. Thus, *Otx*-l and *Otx*-2 are expressed in similar overlapping patterns that are largely restricted to the midbrain and forebrain in midgestation. At E 10. *Otx*-1 is expressed throughout the dorsoventral extent ofit domain that comprises the midbrain and much of the forebrain. The caudal mesencephalic boundary is a ring-shaped transverse band that stops abruptly in front of the fovea isthmi. Following this limit at the m2/Ist boundary is it thin gap, and then expression continues caudally in only the basal plate of the isthinic and cerebellar segments, ending sharply at the cb/r1 boundary. After E10, expression is also found in the entire cortex (p4-p6) with it ventral boundary at the dorsal limit of *Dlx*-1 expression in the lateral ganglionic eminence. It is also expressed in the optic stalk, eye, and posterior preoptic area (POP).

At E10, *Otx*-2 is also expressed through the dorsoventral extent of n12-p4, and there is some expression in alar p5 (primordia of the cortex) and in p6 (rostal cortex, Septum, diagonal band, anterior preoptic area, and optic stalk). Like *Otx-1*, *Otx*-2 expression ends abruptly at the m2/ isthmic boundary, but unlike *Otx-1*, *Otx*-2 is not expressed more caudally. By F12.5, expression decreases in the cortex. Although *Otx*-2 expression is continuous across the alar diencephalon, it hits a peculiar pattern at the p3/p4 boundary (the zona limitans). At this boundary, the ventricular zone is positive, whereas the mantle does not express *Otx*-2. Abutting on both sides of the negative mantle zone is the mantle of p2 and p3, both of which express *Otx*-2. Cells in the p3 mantle express *Otx*-2 and *Dlx*-2; cells in the p2 mantle express *Gbx*-2 and possibly *Nkx*-2.2.

There are two genes that are homologues of the *Drosophila* emply spiracle gene: *Emx-1* and *Emx-2*. The empty spiracle gene regulates development of specific structures in the head of *Drosophila*. The Murine genes are primarily expressed in overlapping patterns in the primordia of the cerebral cortex. Expression of *Emx-2* is also found in it domain that extends from the midbrain into p1, and separately, in the mammillary or tuberomammillary area. Other patches of expression in the diencephalic wall appear to be adjacent to the p1/p2 and p2/p3 interprosomeric boundaries. *Emx-1* and *Emx-2* expression in the cortex appears to be restricted at E10 to p4, but extends at E12.5 to the whole cortex (p4-p6). Emx-l expression may be stronger and extend more laterally in the cortex then *Emx-2*. *Emx-2* is also expressed in the chorioid plexus of the lateral ventricle and the caudal ganglionic eminence.

There are four reported mouse genes and one reported *Xenopus* gene that are homologues of the *Drosophila NK2* gene (the function of the *NK2* gene is not known). The mouse genes are called *Nkx-2.1*, *Nkx-2.2*, *Nkx-2.3*, and *Nkx-2.4*, *Nkx-2.2* is located on mouse chromosome 2. The expression patterns of *Nkx-2.1* and *Nkx-2.2* have been reported. *Nkx-2.1* appears along the basal plate and floor region of p4-p6 at -E10. At E12-5, a separate domain occupies the hemispheric stalk which includes the eminentia thalami, anterior entopeduncular area, and preoptic area as well as in the cerebral hemispheric (in the medial ganglionic eminence and-d lagonal band). At E14.5, expression in EMT diminishes or disappears.

Nkx-2.2 is expressed in a longitudinal strip that overlaps the alar/basal boundary of the whole midbrain and forebrain. Its caudal limit is at the isthmus and it extends to the rostral limit of the brain, where it is just ventral to the optic stalk and suprachiasmatic area. Additionally, *Nkx* is expressed in two alar transverse strips that are immediately caudal to the p2lp3 and p3/p4 boundaries. There is fragmentary evidence that a similar pattern may also be found just caudal to the m/p 1, p4/p5, and p5/ p6 boundaries.

There are at least four murine *Dlx* genes. The sequence of their homeodomains is homologous to the *Drosophila distal-less gene* (*Dll*). Distal-less is expressed in the primordia of the limbs, in ectodermal facial appendages, and in the brain of fly embryos. Mutations of distal-less affect development of the appendages; the effect on brain development has not yet been investigated. *Dlx* genes have been identified in zebrafish, *Xenopus*, newt, and chicken. In the mouse the expression patterns have only been reported for *Dlx*-1 and *Dlx*-2. These genes are expressed in the forebrain, limb ectoderm, and the branchial arches. *Dlx*-1 and *Dlx*-2 are located within about 10 kb of each other on mouse chromosome 2. They are also located on human chromosome 2. Neural expression of mouse *Dlx*-2 begins around E9.5 along an alar longitudinal domain (p3-p6) that spans from the zona limitaxis, through the primordia of the ventral thalamus to the region just caudal to the optic stalk. A few hours later, a second zone of expression begins that extends from the preoptic area along most of the basal telencephalon (p5-p6). The neural expression pattern of *Dlx*-1 is largely indistinguishable with that of *Dlx*-2 (at E 12.5), except that *Dlx*-2 may initiate expression before *Dlx*-l (available evidence suggests that *Dlx*-l expression begins around E10).

The most extensive data come from studies at E12.5 that strongly suggest that these genes are co-expressed in the same cells and that also show that they respect both transverse and longitudinal boundaries in the neuroepithelium. As described later, the study of the *Dlx* genes, in conjunction with the expression of the *Gbx*-2 and *Wnt*-3 genes, stimulated us to propose a neuromeric model of the forebrain. A similar study was performed using *Dlx*-1 and *Wnt*-3.

Expression of the *Dlx*-1 and *Dlx*-2 genes largely turns off postnatally, although several additional features are of note. First, although expression is not found in the dorsal telencephalon at E12.5, expression of these genes is found at later stages in this region. In particular, expression is found in the postnatal subventricular zone of the neocortex at a time when gliogenesis is predominantly occurring.

The mouse *Msx*-1 and *Msx*-2 were originally called *Hox*-7.1 and *Hox*-8.1, and are related to the *Drosophila* Muscle segment homeobox (*Msh*) gene. As described in the next section, they are expressed during development of the eye. Furthermore, *Msx*-1 is expressed in roof plate cells of the neural tube, as well as the neuroepithelial primordia of the chorioid plexus. These genes also have widespread expression in a variety of nonneural tissues of the head. For instance. *Msx*-2 is expressed in cephalic neural crest-derived mesenchyme. This is of note because a mutation of the human *Msx*-2 gene is associated with dominant craniosynostosis.

Homeobox genes expressed in the pituitary

Although it is debatable whether anterior pituitary is derived from the neuroepithelium, for the purposes of this chapter it is included as part of the forebrain because of its close association with the hypothalamus and because of the illuminating work that has elucidated the role of a *POU* protein (*Pit*-1) in its development. Other homeobox genes that are expressed in the embryonic anterior pituitary include *Pax*-6 and *Bmt*-3.0 :*Otx*-2 is expressed in the embryonic posterior pituitary. The anterior pituitary derives from an invagination of the roof of the oral cavity (Rathke's *pou*ch) that comes into contact with a midline evagination of the hypothalamus (the infundibulum). There are five major cell types in the anterior pituitary that secrete peptide hormones: growth hormone, prolactin, adrenocortico-tropin hormone, thyroid-stimulating hormone, and melanotropin-stimulating hormone. In an effort to understand the genetic regulation of the growth hormone gene, Rosenfeld and co workers identified a protein that bound to an enhancer element of this gene. They then cloned this binding protein and thereby identified one of the first *POU* proteins, which they named *Pit*-1. *Pit*-1 is expressed in a restricted domain of the anterior pituitary during embryogenesis, and the onset of its expression occurs before the appearance of cells expressing growth hormone or prolactin. Mutations in this gene, in mice disrupt development of the anterior pituitary, resulting in hypoplasia of the organ. Hypoplasia is caused by the absence of the cells that produce growth hormone, thyroid-stimulating hormone, and prolactin; other cell types still develop. On the other hand, a *Pit*-1 mis-sense mutation in humans resuits in the absence of somatotrophs and lactotrophs without hypoplasia. Thus, both the murine and human mutant *Pit*-1 proteins result in the inability to form specific differentiated cells, while only the mouse mutant has it deficiency in cell proliferation and/or viability. The explanation for this difference is unclear at present, although a hint has come from the discovery that another gene downstream from *pit*-1 is the growth hormone-releasing factor receptor (GRFR) gene. GRF stimulates mitosis in the precursors of growth hormone producing cells.

Thus, a defect in *Pit*-1 can cause less expression of GRFR, and thereby result in an inability of these cells to proliferate in response to GRF. As Rosenfeld and co-workers point out, the regulation of a trophic factor receptor by *Pit*-1 provides a model for the coordination of cell proliferation and the progression of cellular differentiation by *POU* domain proteins.

The neural retina is derived from an evagination of the forebrain, and a number of homeobox genes have been identified that are expressed in this tissue. These include Msx-1, Msx-2, *Pax-6*, *Dlx-1*, *Dlx-2*, *Otx-1*, *Otx2*, *Gtx*, *Brn*-3.0, *Brn*-3.21 *LH*-2, and *Ist*-1. The *Msx* genes, originally called *Hox*-7.1 and *Hox*-8.1, are expressed in distinct compartments within the embryonic eye. *Msx*-2 is expressed in the surface ectoderm and the optic vesicle before invagination occurs. These regions are the primordia of the cornea and retina, respectively. *Msx*-1 is expressed after formation of the optic cup in a domain that becomes the ciliary body. *Pax*-6 is expressed in the primordia of the retina, lens, and the cornea. As noted earlier, mutation of *Pax*-6 results in a defect in lens induction in mice and in the absence of the iris in humans. Analysis of *Dlx*-1 and *Dlx*-2 expression has shown that these genes are expressed in the primordia of the retina. *Otx*-1 and *Otx*-2 are expressed in the optic stalks and vesicles at E10 as well as in a variety of optic structures later in development. The *POU* gene *Brn*-3.0 is transiently expressed in terminally differentiating retinal ganglion cells.

Chapter 6

Homeobox and *pax* Genes

The zebrafish *(Brachydanio rerio)* is an ideal vertebrate for the genetic analysis of development, investigators in several laboratories have begun studying the putative developmental regulatory genes in this organism. The relatively simple features of the translucent zebrafish embryo offer special opportunities for analysing and manipulating the development of single identifiable cells in the central nervous system (CNS) and other tissues.

In several cases it has been possible to analyse embryonic gene expression at higher resolution than has been achieved in other vertebrates. In addition, the evolutionary distance of more than 400 million years between fish and mammals may allow the most essential functions to be identified on the basis of the degree of conservation of gene structure and expression. Most of the putative developmental regulatory genes characterized so far for zebrafish are members of the subfamilies of homeobox genes that have previously been analysed in mammals and *Drosophila*.

In many cases it has been possible to unambiguously identify the fish homologues of specific mammalian genes. Interestingly, the degree of amino acid (aa) sequence conservation between the homologues varies considerably. This relationship has been most extensively analysed for the zebrafish *hox* genes.

Zebrafish counterparts of mammalian homeobox (hox)-containing genes located in both the Hox-B (Hox-2) and Hox-C (Hox-3) clusters have been isolated, but it is still unclear whether the fish genome also contains a cation of the two closely linked Hox-A or a Hox-D complex. The identification of the two closely linked zebrafish genes *hox-B5* (*hox-2.1*) and (*hox-2.2*), which correspond directly to a similar related pair of genes in the mammalian Hox-B complex, confirmed the existence of the Hox-B cluster in the zebrafish genome. Comparison of hox-B5 with its murine homologue revealed several highly conserved features. The putative protein products are of similar length (about 275 residues), and the 96 aa C-terminal, including the 60 aa homeodomain and the pentapeptide, are identical. Within the N-terminal region, the first 25 residues are identical, but the remaining parts show less homology. These features are reflected in a total level of protein sequence conservation of 81%. A significantly lower level (70%) of sequence identity between homologues was observed for the protein derived from the adjacent gene, *hox-B6*. The conservation for this gene is also very high in the homeodomain-containing region and at the N terminus. A similar distribution of homology is present in the *hox-C5 (hox-3.4)* protein. The homeodomain predicted from the zebrafish *hox-C5* gene is identical to the corresponding sequences of the murine *hox-C5* and human *HOXC5* cognates. Further comparison to the human HOXC5 protein showed that the corresponding zebrafish and human proteins are quite diverged. This divergence reflects a smaller amount of conservation in the regions flanking the homeodomain and a particularly low level of sequence identity (18%) in the middle part of the protein.

The differences in divergence of *hox-B5, hox-B6,* and *hox-C5* show that the conservation of the individual genes in the vertebrate Hox clusters is quite variable. Moreover, the data suggest a faster divergence rate for the *Hox*-C genes. Supporting this interpretation, the C-terminal region encoded by the zebrafish *hox-C6 (hox[zf-61]* gene is also quite diverged relative to its murine counterpart. Zebrafish and mammalian homologues located in the Hox-B and Hox-C clusters show very similar embryonic expression patterns in addition to structural conservation.

The most essential elements of the regulatory network involving these Hox genes have probably been maintained during evolution. To identify candidate control regions, noncoding sequences of the zebrafish and mammalian cognates have been compared. Surprisingly, the zebrafish and murine genomic regions, including the *Hox-B5* and *Hox-B6* homologues, contain very few conserved sequences outside the open reading frames. For both homologous pairs, the most significant conservation is present in stretches extending about 0.3 kb upstream from the initiation codons.

In the case of *Hox-B6,* these zebrafish and mouse upstream sequences are 67% identical, whereas the corresponding conservation of the *Hox-B5* homologues is 92%. Although these strong homologies in the upstream regions probably reflect a functional significance in transcriptional and/or translational control, the sequences appear to contain very few recognition elements for known transcription factors.

In contrast, comparison among the noncoding sequences of the zebrafish, murine, and human *Hox-C6* homologues revealed conserved elements that seem to be more directly linked to transcriptional control. Within a 180-bp TATA box-containing region, the zebrafish gene has about 60% sequence identity with its murine and human cognates. The two homeodomain binding sites identified within this promoter region of the human

Hox-C6 gene are also present in the zebrafish sequence. It is not yet clear whether the structural organization of the zebrafish hox clusters correlates closely with the arrangements reported for the four mammalian complexes.

The best evidence for such a correlation concerns the Hox-B cluster. In addition to *hox-B5* and *hox-B6,* a homeobox sequence with striking similarity to *Hox-B4* has been identified.

Using the murine *Hox-B1, Hox-B2* and. *Hox-B3* genes as probes, several cross-hybridizing zebrafish cDNA clones have also been isolated. Despite these correlations, several observations suggest that the zebrafish and mammalian Hox complexes may differ somewhat in structural organization.

In agreement with this idea, a zebrafish sequence *(hox[zf-54])* closely related to Hox-B6 has been identified, which seems not to be a cognate of either of the two other mammalian genes of the same subgroup. Therefore, this zebrafish gene seems to be the result of a unique *Hox-B5* duplication in fish. To clarify this issue, it will be necessary to sequence the remaining part of the *hox[zf-54]* gene and to determine whether or where it is located within the zebrafish *hox* clusters. In the case of the *hox[zf-114]* gene, the correspondence to the mammalian *Hox* genes is even more complicated. Although the entire coding sequence of this gene is known, significant homologies have only been identified for the homeodomain and the pentapeptide.

Among the known mammalian *Hox* sequences, the homeodomain of *hox[zf-114]* is most similar to that of Hox-B3 (79%), but these two genes are definitely not true homologues. There are also differencs indicating that *hox[żf-114]* does not belong to the third group of paralogous Hox genes at all, but rather should be placed in group 5. Alternatively, *hox[zf-114]* could have evolved as a consequence of a separate duplication event after the divergence of fishes and tetrapods, thus being a fish-specific gene.

Although the major features of the *Hox* clusters in fish and mammals are likely to be very similar, further analysis of the zebrafish hox genes could provide additional examples of divergence that may help to elucidate the evolution of this vertebrate gene family. Similar to the studies of hox genes, analyses of zebrafish *engrailed* enrelated genes have revealed several features that diverge from those of other vertebrates.

Apart from zebrafish, only two genes of the engrailed type have been identified in the vertebrate species analysed so far. A potential zebrafish homologue of the mammalian *En-2* gene was reported by Fjose and colleagues, but this study suggested, based on Southern analysis, that the zebrafish genome might contain more than two *engrailed sequences*. This proposal has been confirmed by Ekker and colleagues, who have identified and characterized three distinct *engrailed genes (eng-l,* eng-2, and *eng-3).*

Whether the zebrafish has additional *engrailed-like sequences* is still unclear, but the cross-hybridization analysis indicates the existence of at least four members of this gene family. Multiple engrailed-related genes may not be a unique feature of fish. The finding of a third, and highly diverged, engrailed-like gene in *Drosophila* provides some support for this assumption. Differences between zebrafish and mammalian *engrailed genes* are also clearly evident at the level of protein sequence homology. Although the putative cognates contain several highly conserved domains, the total level of sequence identity between the zebrafish and murine proteins is quite low

(65-69%). Among the three known zebrafish engrailed sequences, eng-l is the most closely related to the murine *En-1* gene. The conservation of the putative Eng-1 protein, which shares 65% sequence identity with En-1, is mainly located within the five *engrailed homology* regions (EHI-5) that are common to all known engrailed genes. Thus, apart sequence conservation is present between the N-terminal domains of Eng-1 and the murine En-l protein. Several characteristics indicate that *eng-2* and *eng-3* are products of a duplication of an ancestral En-2 gene.

The putative protein products of *eng-2 and eng-3* share 65 and 69% sequence identity, respectively, with the murine En-2 protein. The predicted zebrafish proteins have extended homologies in the EH I and EH5 regions relative to En-2.

An additional serine-rich peptide sequence, which was initially found to be conserved in the N-terminal part of Eng-2 and *Xenopus* XEn-2, is also present in the proteins derived from zebrafish *eng-3 and* the mouse *En-2* gene. Moreover, these four genes have conserved sequences in the 3' untranslated region. In addition to the *engrailed genes,* more than 15 distinct subgroups of vertebrate homeobox genes have been defined on the basis of differences among their derived homeodomain sequences, and many of these groups are related to specific *Drosophila* genes. Zebrafish genes that belong to at least six of these subgroups have so far been isolated. When the zebrafish members of each of these groups are compared to the corresponding mammalian genes, the degree of sequence identity is quite variable, perhaps because of differences in functional conservation. Although four zebrafish genes containing muscle *segment homeobox (msh)* type sequences have been characterized, their evolutionary relation ship to the two known murine members of this subgroup, *Msx-l* (*Hox-7*) and *Msx-2*

(*Hox-8*), remains to be determined. Comparison of the putative Msx-l/Msx-2 proteins with the aa sequences derived from the zebrafish *msh*-C and *msh-D* genes revealed a high degree of sequence identity (>90%) within an 88 amino acid region containing the homeodomain. However, outside this conserved region, there is very little sequence similarity among any of the four proteins. Thus, it is not possible to identify true homologues from sequence comparison alone, and the current information about the expression patterns is not sufficient to resolve this issue.

For the zebrafish *msh* A and *msh-B genes,* where only the homeobox sequences have been determined, further sequencing is required to see how they relate to the two murine Msx genes. However, it also remains to be determined whether the zebrafish and mouse genomes contain the same number of *msh*-like genes. Since more genes of this subgroup have been identified for zebrafish, it could be that independent duplications have occurred in mammals and fish.

For the zebrafish gene *cdx[Zf-cadl],* which belongs to the caudal group, the extent of homology relative to the known members in mouse, chicken and *Xenopus is* quite limited. The degree of sequence identity between the different homeodomains is moderate (80-86%,), and only a few short stretches of conserved amino acids are present in the N-terminal region. Thus, for none of these vertebrate species have true homologues been identified. Further research is clearly required to determine how many caudal-like genes are present in these different vertebrate genomes and how these sequences are related. The putative protein products of the zebrafish, *Xenopus,* and murine cognates of the *goosecoid gene* are very similar to the Hox proteins with respect to the distribution of conserved sequences. While a C-terminal region of about 100 amino acids that includes the homeodomain is highly conserved (>93%), a moderate level of sequence identity (<65%) is present in the N-terminal parts. us proteins have an overall identity of about 75%. Whether the genes also share a significant degree of sequence identity in noncoding; regions has not been determined. Other subgroups of homeobox genes for which zebrafish members have been identified include *Distal-less* homeobox *(d/x),* H2.0-like homeobox *(h/x),* and *Pit/0ct/Unc*(POU) type sequences. Since only limited sequence data have been obtained for the zebrafish genes, it is premature to draw any conclusions regarding their conservation.

Zebrafish homologues of several other categories of vertebrate developmental genes, including the *wingless homologue int-1* (*Wnt-l*) proto-oncogene, *Kruppel*-related zinc finger (*Krox-20*), chicken ovalbumine upstream promoter (COUP) transcription factor, and the *Brachyury* (*T*) gene, have been reported. The fish and mammalian proteins derived from these genes have a level of sequence identity (69-76%) comparable to that of the Hox and *eng* cognates. The exception is the zebrafish COUP homologue, *sup[44],* which displays the same profound evolutionary conservation as *pax[zf-a].*

Eight *paired box-containing* genes (Pax) that are known for the mouse, four also encode *paired* type homeodomains. For one of these genes, Pax-6, a true homologue (*pax[a]*) has been identified in zebrafish. In contrast to the known *hox* and *eng* cognates in zebrafish, where only the homeoboxes and other discrete elements show it high degree of sequence similarity, the entire protein-coding regions of the two Pax-6 homologues are highly conserved. Both species also have a variant form of Pax-6 transcript that includes an identical insert of 14 amino acids in the paired domain. The functional importance of the insert is presently unclear. However, its location within the paired domain is likely to affect

DNA binding and/or protein-protein interactions and thereby provide an additional mode for developmental regulation. Another zebrafish pax gene, *pax[b]*, is highly conserved relative to three murine genes *(Pax-2, Pax-5, Pax-8)* that form a subgroup of Pax genes. The putative protein products of *Pax-2, Pax-5,* and *Pax-8* have a total similarity of 87, 71, and 65%, respectively, to the predicted amino acid sequence of *Pax[b]*. Thus, *pax[b]* and *Pax-2* seem to be true homologues. Potential zebrafish homologues of *Pax-1, Pax-3,* and *Pax-7* have also been identified, but the entire coding sequences have not yet been determined. At present it is unclear whether the zebrafish genome contains additional members of the *Pax-2/Pax-5/Pax-8* subgroup.

The spinal cord and hindbrain regions of the embryonic CNS have several features in common, indicating that the underlying developmental regulatory mechanisms are similar. Because these are general properties of all vertebrates, any species could in principle be used as a model to investigate these mechanisms. However, in this context, the relatively simple organization of the zebrafish hindbrain and spinal cord provide several advantages in analysing the genetic networks controlling neural patterning. The hindbrain of zebrafish embryos becomes transiently subdivided into a series of rhombomere segments that are morphologically visible. Moreover, the seven most-anterior rhombomeres seem to be directly homologous in the different vertebrate species. In chicken, the morphologically visible rhombomeres are defined as segmental units on the basis of several criteria, including lineage restriction, mitotic patterning, arrangement of somatic motor nuclei, and the localization of transverse axons at the boundaries. Some of these criteria have been confirmed for the zebrafish rhombomeres, and a segmental arrangement is observed at the level of single identified neurons.

The reticulospinal interneurons can be assigned to different families of segmental homologues on the basis of morphological similarities. Consistent with the idea that rhombomeres are of developmental significance, several categories of regulatory genes have rhombomere-restricted expression.

The various genes within the individual *Hox* clusters have overlapping expression domains, and the anterior expression limit of each gene correlates closely with the position of the gene in the complex. In the hindbrain, the 3' members of a particular complex are expressed in segment-specific patterns, with successive genes having borders at two-segment intervals. This pattern suggests an involvement in segment formation or specification of segment identities, and similar correlations would be expected for the neural expression of zebrafish *Hox* genes. In agreement with this prediction, analysis of the zebrafish *Hox-B5-6* and Hox-C5 cognates revealed that the anterior limits of expression are located in the region of junction between the hindbrain and spinal cord. However, the limited resolution achieved in these experiments, together with the lack of distinct segmental boundaries in the posterior hindbrain, prevented direct comparison of these homologues in mouse and zebrafish. Thus, it has not yet been directly demonstrated that the 3' Hox genes of zebrafish follow the same rules of collinearity and rhombomere-restricted expression as their murine counterparts.

Indirect evidence for such a relationship has been obtained from studies on the zebrafish homologue *(krx-20)* of the murine *Krox-20* gene, which encodes a protein with three C_2H_2-type zinc fingers. In mouse embryos, *Krox-20* is expressed at a high level in rhombomeres 3 and 5 (r3 and r5), both before and after they become visible. Accordingly, it has been postulated to be a segmentation gene regulating *Hox* expression, and direct evidence for this hypothesis was obtained for the expression of *Hox-B2*. In zebrafish, *krx-20* transcripts are first detected in the primordia of r3 and r5 at 100% epiboly (9-10 hr), suggesting that segmentation in the zebrafish hindbrain, at the molecular level, occurs at this stage. This is several hours before the morphological manifestation of rhombomeres in 15- to 16-hr embryos, and it could therefore regulate the segment-restricted expression of zebrafish *hox* genes. Another example of segmental expression in the zebrafish hindbrain is the *h/x-1* gene, which belongs to the same group of homeobox genes as H2.0 in *Drosophila*. In this case, a transverse stripe of expression was first observed in the r5 primordium of 12-hr embryos. At a later developmental stage (15-16 hr), different levels of h/x-1 transcripts were also detected within rhombomeres 1-3. Because the expression domains are dorsoventrally restricted, *h/x-1* is probably not involved in segmentation. Moreover, the level of expression varies for the individual rhombomeres and could be regulated by factors such as Hox proteins, which are likely to specify segmental identities.

Studies of the embryonic expression of *pax[b]* and the zebrafish *eng* genes have revealed aspects of segmentation and AP specification at the level of single cells. In both cases, the expression patterns have been investigated by *in situ* hybridization and immunohistochemical methods. However, for the *eng* genes, the results obtained with the two different methods are somewhat contradictory. In the hindbrain, the 4D9 antibody, which cross-reacts with all of the three known zebrafish Eng proteins, labels clusters of cells only in the three most-anterior rhombomeres. The other antibody, originally generated against the mouse EN-2 homeodomain, aEnhb-1, revealed a more complex expression pattern in which clusters of stained cells were detected in all of the rhombomeres and the spinal cord.

The problem of identifying which of the *eng* genes are expressed in the hindbrain and spinal cord cells, several interesting conclusions can be derived from these studies.

The location of the expressing cells suggests that they consist of at least two different types of interneurons. Cells labelled with αEnhb-1 are detected earlier by (18 hr) than the 4D9-positive cells (24 hr) and are not restricted to a specific AP level. The segmentally repeating αEnhb-1 pattern also appears transiently in the spinal cord, where expression extends posteriorly over time. While the younger hemisegments have one or two labelled cells, the older segments (more anterior) show expression in additional cells that fill in the row until the periodicity is lost. *Engrailed*-expressing cells in the hindbrain and spinal cord are partially regulated by the same segmentally repeating information, even though most of the spinal cord neurons are not segmentally organized. The hindbrain rhombomeres exhibit unique segmental identities, with each segment containing a specific set of identifiable reticulospinal neurons. Results obtained from studies on the *pax[b]* gene are compatible with this model and demonstrate the advantage of using the zebrafish for analysing these aspects at the molecular level. A polyclonal antibody raised against the C-terminal end of the Pax[b] protein (which does not include the conserved paired domain) was used to investigate expression in zebrafish embryos. At the 13-hr stage, Pax[b] staining was first detectable in a distinct set of 12-15 nuclei on either side of the neural keel in the region including the posterior hindbrain and anterior spinal cord. The Pax[b]-positive cells in the hindbrain of 21-hr embryos were identified by their location and by axonal trajectories made visible by double labelling with acetylated α-tubulin antibodies. According to this analysis, the *pax[b]* gene is expressed in the MiD2c primary commissural interneuron in r5 and in a pair of similar interneurons (MiD3c) in r6. In addition, a previously unclassified set of interneurons that was termed commissural Caudal-rhombomere ascending (CoCaA) was shown to be Pax[b] positive in the caudal hindbrain. In the spinal cord, where the *pax[b]* expression pattern does not have segmental features, the Pax[b]-positive neurons have also been identified. These observations suggest that *pax[b] is* in some way involved in determining the specific fates of particular neurons or their subsequent differentiation. In the case of the MiD2c and MiD3c cells, Pax[b] proteins are detectable 3 hr before these neurons start to project their axons. This timing of pax[b] expression is consistent with a role in determining cell fate, since some neurons in zebrafish embryos are unalterably specified as late as 1 hr before they project their growth cones.

Using an antibody raised against the *Xenopus* homeodomain protein XlHbox1, which probably is the homologue of *Hox-C6*, expression (commencing at the 16-hr stage) was detected in the anterior spinal cord. Within this region, a dorsal row of X1Hbox1-positive cells were identified as Rohon-Beard (RB) sensory neurons. Interestingly, at 25 hr these neurons exhibit a graded decrease in staining intensity from anterior to posterior over the domain of spatiotemporal expression pattern indicates that the zebrafish Hox-C6 homologue is involved in AP specification of these neurons. However, the gene probably does not play a role in specifying the RB cell type, since expression is first detected after these neurons start to differentiate.

The spinal cord of vertebrates exhibits a characteristic dorsoventral (DV) pattern, with specific classes of neurons differentiating according to their DV position. Moreover, surgical manipulations of chicken embryos have shown that the notochord plays a crucial role in ventralizing the neural tube by inducing cells at the ventral midline to become floor plate cells. Studies indicate that the *Pax* genes are of particular importance in controlling the embryonic DV patterning.

The murine *Pax-3* and *Pax-7* genes are expressed in partially overlapping areas in the dorsal region of the ventricular zone. *Pax-7* is expressed in the alar plate region, excluding the roof plate and neural crest cells, while *Pax-3* expression is detected in all three of these subregions. Similarly, the murine *Pax-6* gene is active in the entire basal plate, excluding the floor plate and cells directly adjacent to it. Analyses of the embryonic expression of the zebrafish *Pax-6* homologue, *pax[a]*, have revealed the same kind of DV restriction. In addition, preliminary data from *in situ* hybridization analyses of the potential zebrafish homologues of *Pax-3* and *Pax-7* were consistent with the observations for the two murine genes. These results indicate an involvement of *Pax* genes in specifying DV identities of different populations of mitotic cells in the ventricular zone.

Pax genes may also be involved at later stages of neuronal differentiation. Consistent with this idea, the Pax-2, Pax-5, and Pax-8 genes are expressed in postmitotic cells located at specific levels in the intermediate zone in the spinal cord of mouse embryos. Although these cells are probably precursors of different types of interneurons, it has not been possible to determine their exact identities. In contrast, the relatively simple organisation of the zebrafish spinal cord at early embryonic stages makes it a convenient model for analysing neuron-specific expression patterns.

A small number of identifiable classes of neurons, including sensory RB cells, motor neurons, and several classes of interneurons, are present in 18- to 20-hr embryos. By exploiting this knowledge, it was possible to determine the identity of spinal cord neurons expressing the zebrafish Pax-2 homologue, pax[b]. This identification, which in part relied on double labelling with antibodies recognizing specific neural antigens, showed that the pax[b] gene is expressed in commissural secondary ascending (CoSA) interneurons. These cells are present at variable densities (one to five per spinal hemisegment) along the entire spinal cord. As observed for the pax[b]-expressing reticulospinal neurons in the hindbrain, the CoSA neurons project their axons several hours after the Pax[b] protein is detected. Accordingly, the differentiation of these neurons could also in part be determined by the *pax[b]* gene. Dorsoventrally restricted expression in the embryonic hindbrain and spinal cord has been observed for the murine *Dbx* and chicken *ChoxE* genes, which belong to the same group of homeobox sequences as the zebrafish *h/x-1* gene. In both cases, bilateral columns of expression, restricted to the proliferating zone in the dorsal basal plate, extend throughout the hindbrain and spinal cord. For *h/x-1*, a, similar expression pattern was observed, but in this case more detailed information was obtained by whole-mount *in situ* hybridization analysis. After a segmental *h/x-1* pattern appears in the hindbrain of 12- to 15-hr embryos, expression extends as a contiguous column throughout the hindbrain and spinal cord. This dorsoventrally restricted column, which is located near the dorsal boundary of the basal plate, may include both mitotic and postmitotic precursors of particular types of neurons.

Analyses of the neural pattern of rhombomeres indicate that the individual segmental units can be subdivided along the AP axis. In the hindbrain of the chick, motor nuclei are in the segment centers, and early commissures are near the borders. Also, the centers of rhombomeres have a higher mitotic density and shorter cell-cycle time than the boundaries.

Expression of the neural-cell adhesion molecule N-CAM is localized to the central regions of the rhombomeres, whereas Ng-CAM/L1 is restricted to the boundaries. Despite this patterning of individual rhombomeres, none of the developmental regulatory genes analysed so far is expressed selectively within particular subregions.

In zebrafish, these aspects have been investigated in further detail with various monoclonal antibodies that stain different subpopulations of cells within the hindbrain. This revealed an arrangement of neurons, commissural tracts, neuropil areas, and radial glial fibers that reflects an alternating pattern along a series of seven segments. In the model proposed from this pattern, the rhombomere centers contain the first basal-plate neurons to develop and the first neuropil. The other region, surrounding the segment boundaries, contains the first alar-plate neurons. Studies of the embryonic expression of the *h/x-1* gene revealed spatiotemporal changes in the hindbrain of 1-day-old embryos that seem to be consistent with this model. Each longitudinal Column of expressing cells transforms into series of transverse stripes, and by the 30-hr stage a repeated pattern of paired stripes is clearly visible along both sides of the midline. The spacing between each pair of *h1x-1* stripes correlates well with the width of hindbrain rhombomeres.

Expression studies of various types of vertebrate developmental genes in the hindbrain and spinal cord have revealed clear correlations with the neural patterning observed along the AP and DV axes. Because of the great complexity of the rostral brain in mouse embryos, a relationship between the expression domains of regulatory genes and morphological subdivisions appears to be harder to elucidate. Studies on *pax* and *eng* genes have shown that the relatively simple organization of the zebrafish brain at early embryonic stages may facilitate analysis of the genetic mechanisms responsible for its morphogenesis. Whole-mount, in situ hybridization analysis of the zebrafish *pax[b]* gene revealed the earliest signs of regionalization in the rostral neurectoderm of 9-hr embryos. At this late gastrula stage, *pax[b]* transcripts are detectable within two transverse bands of cells separated by a nonexpressing area at the dorsal midline. Both the width and the length of these stripes, which point somewhat posteriorly at an angle of about 60° relative to the midline, increase simultaneously with the expression level during the subsequent hour of development.

As a result of convergence of cells that generate the neural keel, the two stripes start to fuse at the dorsal midline in 10-hr embryos. Thus, by the 12-hr stage, the *pax[b]* expression domain is a solid transverse band in the neural keel. Analysis of later developmental stages showed that this area corresponds to a region including the midbrain-hindbrain boundary. Pax[b] proteins were detected in 9-hr embryos using the *pax[b]* antibody, and the spatiotemporal pattern in subsequent stages appeared to be the same as for the transcripts. In the case of *eng*-2 and *pax[a]*, the observed spatiotemporal expression patterns are similar to those for *pax[b]*. However, while the transverse stripes of *eng*-2 expression are present at the same AP level as *pax[b]*, the *pax[a]* gene is active in two separate domains, located in the forebrain and the hindbrain regions.

In the anterior hindbrain, pax[a] is initially expressed within two transverse stripes that later are extended posteriorly throughout the hindbrain and the spinal cord. The rostral domain of *pax[a]* expression first appears as two triangular areas that fuse at the midline. For both of the *pax[a]* domains, the transverse expression boundaries are almost parallel to the *pax[b]* and *eng*-2 stripes. This pattern is reminiscent of the segmental expression domains of their *Drososphila* homologues *paired* and *engrailed* and suggests the involvement of related subdividing mechanisms in the midbrain region. It is interesting to note that the *pax[b]/eng-2 stripes* are located within the gap between the two *pax[a]* domains. The expression of *eng-l* and *eng-3,* which has not yet been analysed at this early stage, is likely to be restricted to transverse stripes within this gap. Whether these relationships reflect repressor functions for *par[b]* and the three *engrailed* genes in the regulation of pax[a] transcription remains to be demonstrated.

A periodic array of swellings, called neuromeres, is present in the neural tube during early embryonic stages of zebrafish and other vertebrates. Rostral to the seven segment-like hindbrain neuromeres (rhombomeres), three distinct neuromeres have been identified in zebrafish embryos. These neuromeres, which become visible in 16-hr embryos correspond to the definitive midbrain and the two divisions of the forebrain, the diencephalon and telencephalon.

Although the rostral brain neuromeres are about three times as long as the hindbrain rhombomeres, they have some related features of organization. Similar to the bilateral clusters of early differentiating neurons which appear in the basal plate of individual rhombomeres, two clusters positive for acetylcholi-nesterase (AChE) have been identified in the basal plate of the midbrain and diencephalon. However, reflecting their larger size, the initial clusters in the rostral neuromeres contain more cells. Like the rhombomeres, the midbrain and diencephalon are separated by a commissure. This prominent posterior commissure (PC) crosses dorsally, whereas the intersegmental commissures in the hindbrain are located ventrally. In contrast to the diencephalon, midbrain, and hindbrain neuromeres, the first neural cluster of the telencephalon appears in the alar plate.

Thus, on the basis of morphological criteria, it remains unclear whether the rostral neuromeres are segmental units.

Analysis of the expression patterns of the genes that are regionally expressed in the rostral brain may help to resolve this issue. Several characteristics of the *pax[a]* expression pattern appear to link this gene to regionalization in the rostral brain. A restricted domain of expression is first detected at the 9- to 10-hr stage, more than 5 hr before the neuromeres are formed. Also, as expected for a regionalizing gene, the transcripts are uniformly distributed in all cells within the area of expression. In addition, the posterior expression border coincides well with the boundary between the diencephalon and midbrain neuromeres that has become visible at the 17-hr stage.

The area of expression remains almost unaltered during the period when the first morphological subdivisions and neuronal differentiation occur.

Within the forebrain, the *pax[a]* expression domain is restricted to the presumptive thalamic part of the diencephalon. Therefore, the gene is likely to play it role in specifying this subdivision of the diencephalon. Surprisingly, preliminary studies of the phenotype of mice carrying mutant alleles of the homologous Pax-6 gene revealed no defects in the forebrain. One possible explanation for this finding is that subtle alterations were not detected, due to a lack of appropriate markers. Alternatively, the normal phenotype could reflect functional redundancy of Pax genes with overlapping expression domains. It is tempting to speculate that the stripes of pax[b] and *eng*-2 expression observed at early embryonic stages correlate with segment-like units. A double labelling experiment with pax[b] and the HNK-1 antibody, which labels the hindbrain rhombomeres r2 and r4, has revealed some evidence in support of this proposal. In 14-hr embryos, a periodic pattern was observed, in which a transverse stripe of pax[b] expression with a width similar to that of a rhombomere was located approximately one segmental unit anterior to the HNK-l labelling in r2.

Results obtained from grafting experiments in chickens, and targeted disruption of the *En-2* and *Wnt-1* genes in mice have provided strong evidence that these two vertebrate homologues of the segmentation genes *engrailed* and *wingless* in *Drosophila* are required for normal development of the regions flanking the midbrain-hindbrain boundary.

Studies of the corresponding genes in zebrafish have provided more detailed information about the exact relation between the expression domains and the morphological boundary separating the midbrain and hindbrain. The segment-like expression of pax[b] in the same region indicates an involvement in the formation of this boundary. It is interesting in light of this observation that the related *Drosophila* paired box genes, *paired* and *gooseberry*, are required for proper expression of *engrailed* and *wingless* during segmentation. A similar situation could exist in zebrafish, where *pax[b]* expression is initiated earlier than the *eng-2* and *wnt-l* domains and partially overlaps them.

Further evidence for regulatory interactions among these three genes has been obtained from intracellular microinjection experiments with a Pax[b]-spccific antibody. Embryos injected with this antibody shortly after fertilization did not form the furrow at the mid brain-hind brain boundary. This abnormality occurred even though the development of other tissues where *pax[b]* is expressed appeared to be normal. To investigate the molecular mechanisms underlying these observations,

Krauss and colleagues analysed the expression of pax[b], *eng-2,* and *wnt-1* by *in situ* hybridization.

The expression of all three genes was greatly reduced within the transverse midbrain-hindbrain stripes, while other domains were unaffected. These alterations suggest an autoregulatory function for the Pax[b] protein in regulating transcription of the *pax[b]* gene. Moreover, the downregulation of *eng*-2 and *wnt-1* provides experimental support for a regulatory network involving these three genes in the vertebrate embryo. Although these results indicate a key role for *pax[b]* in regulating the formation of the midbrain-hindbrain boundary, it remains unclear how the position of the furrow is determined. In this connection, it is of interest to consider the exact location of the expression borders of the individual genes found to be active in this region of the neural tube. As judged from the expression pattern of the *eng-l* gene, this gene may be of particular importance in defining the location of the furrow.

The anterior border of resolution, but it seems to coincide boundary. The posterior border of the transverse *wnt-1* stripe is located close to the center of this furrow as well. Thus, the *eng-l* and *wnt-l* domains may be directly juxtaposed in the same way as the stripes of wingless and engrailed expression observed at the parasegmental boundaries in Drosophila. This resemblance suggests important roles for *eng-1* and *wnt-1* in specifying the exact location of the midbrain-hindbrain furrow. The eye primordia of zebrafish are first detected as lateral swellings in the diencephalon of 11- to 12-hr embryos. The generation of these primordia is therefore one of the first morphological consequences of the underlying molecular regionalization of the neural tube. Studies of the embryonic expression patterns of *pax[a]* and *pax[b]* have provided evidence that regionalization within the eye primordium also occurs very early. Transcripts of *pax[b]* are first seen at the 12-hr stage, and expression *is* restricted to the region where the optic stalk appears. In 18-hr embryos, the expression level is higher, and the area of expression includes the optic stalk and the ventral part of the optic vesicle. This spatial pattern is also maintained in later developmental stages.

A similar spatiotemporal expression pattern was observed for the murine homologue Pax-2, suggesting that the gene performs a role in eye development. Stronger evidence for a function in the specification of the vertebrate eye has been obtained for pax[a] and its highly conserved murine homologue *Pax-6*. In the case of *pax[a]*, transcripts are first detected in the optic vesicles of 12-hr embryos.

The initial expression is restricted to the portion of the eye primordium that faces the diencephalon excluding most of the optic stalk region. This expression domain expands, and in 24-hr embryos a uniform transcript level is observed throughout the optic cup. At this stage, *pax[a]* transcripts are also detected in the ectoderm overlying the retina, indicatme an involvement in inducing the lens and or the cornea. An interesting feature in this context is that *pax[a]* is expressed in both the inducing (optic vesicle) and the responding tissue. The same result have been obtained for the murine Pax-6 gene and mutant alleles of this gene have been identified that cause abnormal eye development. In homozygotes, eyes do not develop whereas in individuals carrying a single Sey allele, eyes are smaller than normal and have vacuolated lenses. These semidominant features probably reflect a threshold requirement for the Pax-6 protein during the inductive interactions between the lens placode and the optic cup.

A prominent feature of the first neuronal clusters of the rostral neuromeres in zebrafish embryos is that they project axons in a stereotyped pattern and thereby establish an early invariant scaffold of tracts. Axonogenesis occurs in a specific temporal sequence during the 16- to 24-hr stages of embryonic development and therefore correlates with the period when *pax[a]* and *pax[b]* are expressed within sharply defined areas in the rostral brain. A comparison of the axonal scaffold and the expression pattern suggests that several of the early tracts follow pathways along the expression borders of the two pax genes. The reported positions of the tracts of the postoptic commissure (TPOC) and the PC appear to be at the ventral and caudal expression borders of pax[a], respectively.

Anteriorly, the TPOC lies near the border between the pax[b]-expressing optic stalk and the forebrain. In addition, the anterior commissure axons and supraoptic tract seem to follow pathways in the forebrain that are closely juxtaposed to pax[b]-expressing cells in the optic stalk. In the posterior midbrain, the medial longitudinal fascicle, which is pioneered by neurons in the ventral midbrain cluster, also has a pathway near the ventral border of pax[b] expression.

The neural expression patterns of many different homeobox and pax genes have been extensively analysed in zebrafish embryos. Considerably less is known about the genes' mesodermal expression. In the case of *hox* genes, this is due to limitations in the sensitivity of the in situ hybridization method and the fact that the mesodermal transcript levels of these genes appear to be lower than in the CNS. However, recent improvements of the *in situ* hybridization techniques used for zebrafish and the application of immunohistochemical methods will undoubtedly facilitate future investigations of mesodermal patterning and differentiation in this species.

For the zebrafish *hox-C6* gene, the limited sensitivity of the *in situ* hybridization method was avoided by applying a cross-reacting antibody raised against the protein product of the *Xenopus* homologue XlHbox1. With this antibody, weak staining was first detected in anterior somites of 14-hr zebrafish embryos.

By the 16-hr stage, more prominent staining was observed in the somitic mesoderm and expression was restricted to the lateral surfaces of somites 5 to 7. The anterior border of expression coincided exactly with the boundary between somite 4 and 5, while in a posterior direction the expression decreased in a graded fashion over somites 6 and 7. The XlHboxl staining detected in *Xenopus* and mouse embryos was similar, but the correlations between the expression border and somite boundary were defined less precisely. These results suggest a conserved role of hox-C6 in specifying the AP position of somites. Studies on the zebrafish *eng* genes have provided evidence for an involvement in patterning of the individual somites. Using the 4D9 antibody, Eng proteins were observed in clusters of cells located at the medial surface of the somites, facing the notochord. This expression was detected after the 13-hr stage and followed a spatiotemporal sequence that directly correlated with the posteriorly directed wave of somite formation. Thus, for a particular somite, Eng staining appeared 1 to 2 hr after its formation.

The Eng-labelled cells were identified as muscle pioneers, which are the first cells to differentiate within the myotome that develops from each somite. The muscle pioneers undergo extensive morphogenetic changes, and a furrow appears where the horizontal myoseptum eventually forms and separates the dorsal and ventral muscles. On the basis of the correlation between *eng* expression and the generation of furrows both at the midbrain-hindbrain boundary and in the myotome, it was proposed that one or more of the *eng* genes are critical in specifying the positions of these morphological boundaries. The roles of the individual *eng* genes in the myotomes have been further investigated by *in situ* hybridization analyses with gene-specific probes. Expression of both *eng-1* and *eng-2*, but not *eng-3*, was observed in the muscle pioneers. Since *eng-2 is* only transiently expressed, the *eng-l* gene is probably more essential in subdividing the myotomes.

Reflecting the divergence of the *eng* genes, the murine and chicken homologues have somewhat different somitic expression patterns. However, the *En-l* gene is expressed in a lateral stripe in the middle of the dermatomes of both chicken and mouse embryos, indicating a similar role in subdividing the somitic mesoderm. Both whole-mount *in situ* hybridization and immunohistochemical staining have shown that the embryonic kidney is a major site of *pax[b]* expression. Posterior to the position of the second somite, a continuous column of two to three cells expressing *pax[b]* forms a half circle at the edge of the constricting embryonic shield in 13-hr embryos. These mesodermal cells, which include the nephritic primordium and still express the *pax[b]* gene at the 24-hr stage. This spatiotemporal pattern suggests an early involvement in specification and/or differentiation of the cells generating the kidney. Consistent with this assumption, the murine homologue *Pax-2* has a very similar expression pattern, and deregulation of this gene in transgenic mice causes severe kidney abnormalities.

Studies of the embryonic expression of zebrafish hox genes have also revealed evidence for functional conservation in nonsomitic mesoderm. Using the XlHboxl antibody to investigate expression of the zebra fish *Hox-C6* gene, Molven and colleagues observed labelling of the pectoral fin bud. Expression was first detected in the fin-forming region, about 10 hr before the fin bud itself becomes visible in 29-hr embryos as a condensation of lateral mesodermal cells at the level of somites 2 and 3. During this period, the expression becomes gradually restricted to the anterior part, and at 48 hr staining is maximal in the anterior and proximal region of the fin mesoderm. Immunohistochemical analyses with the XlHboxl antibody on Xenopus and mouse embryos revealed very similar expression patterns in the forelimb bud. Thus, these studies have provided some molecular evidence supporting the assumption that the tetrapod forelimb and pectoral fin of early bony fishes have a common evolutionary origin.

Additional evidence for conservation of the genetic mechanisms underlying patterning of the forelimb has been gained from analyses of the eng genes. Engrailed proteins were detected with the 4D9 antibody in the epidermal cells of the pectoral fin bud of 26-hr embryos. This uniform labelling was limited to the ventral-anterior half of the fin bud, and the boundary of expression at later stages followed the ridge of the flattened fin, which is the equivalent of the apical ectodermal ridge (AER) of other vertebrates. Only the *eng-l gene* is expressed in the pectoral fin bud. Similarly, *En-l* expression was observed in the ventral ectoderm of the forelimb bud in mouse embryos. Also in this case the expression boundary coincided with the AER. These observations indicate a conserved role for the *eng-I/En-l genes* in defining differences between the dorsal and ventral regions of the bud and perhaps in specifying the position where the AER forms. In relation to these questions, it will be of interest to investigate whether cell lineage restrictions correlate with the compartment-like boundary.

In addition to the muscle pioneers, eng expression has been detected in mesenchymal cells that develop to form two specific jaw muscles in the zebrafish.

Expression wits initially observed in scattered mesenchymal cells, caudal to the eyes and lateral to the hindbrain in 24-hr embryos.

In later developmental stages, the labelled cells moved rostrally and aggregated to form a single cluster in the mandibular arch, which eventually differentiated as two specific jaw Muscles. Only transcripts of the *eng*-2 and *eng*-3 genes were detected in the muscle precursor cells by *in* situ hybridization.

Interestingly, the growth cones of trigeminal motoneurons were associated with the eng-positive mesenchymal cells a few hours after onset of expression. These observations suggested that the eng genes play a role in specifying a particular subset of muscles from the pharyngeal arch mesoderm and that they indirectly contribute to neuromuscular target recognition. These functions may be conserved, since the murine *En* genes are expressed in mandibular myoblasts. The close similarities between the observed expression patterns of homologous pax and homeobox genes in zebrafish and other vertebrates suggest that the essential mechanisms responsible for regulating embryonic expression are highly conserved. These signalling pathways have been most extensively studied in *Xenopus, chicken,* and mouse embryos, and have provided preliminary evidence for the involvement of retinoids and several types of secreted proteins and their receptors. In the few cases where these aspects have also been investigated in zebrafish, the results are consistent with the findings in higher vertebrates.

Two zebrafish homologues of *Xenopus* and mouse genes that are known to be involved in axial patterning during gastrulation have been analysed in considerable detail. One of these, Zf-T, is directly homologous to the mouse *Brachyury (T)* gene, which is required for normal

AP patterning of the mesoderm and the generation of a notochord. As in mice, the T gene in zebrafish is expressed in the notochord and early mesoderm. Both in *Xenopus* and zebrafish, transcription of the T gene can be activated by the mesodermal-inducing growth factor activin A. This observation suggests that the fundamental inductive events have been conserved.

The *Xenopits* homeobox gene goosecoid can also be induced by activin in the Spemann's organizer region, and the gene probably plays a central role in both DV and AP patterning. Results obtained from analyses of the embryonic expression of the zebrafish *goosecoid* homologue are in agreement with this proposal.

Zygotic expression of the zebrafish gene is first detected at the midblastula stage in the presumptive dorsal lip region. In gastrulating embryos, expression is localized within the anterior axial hypoblast, suggesting that *goosecoid* might be specifically associated with the head organizer, as has been proposed for the Xenopus homologue. One of the immediate consequences of this expression in the anterior hypoblast could be the induction of genes involved in head patterning. Consistent with this assumption, it requirement for vertical signals from underlying axial mesoderm in patterning of the forebrain has been demonstrated in *Xenopus.* This gene could be located downstream of *goosecoid* in a regulatory cascade. At the 8-hr stage, *h/x-1* is induced within a circular area in the rostral region, and following dynamic alterations of the expression domain, the transcripts become restricted within prechordal plate cells underlying the midbrain and forebrain. Further studies are required to determine whether this spatiotemporal pattern reflects an involvement of the *h/x-l* gene in development of the rostral brain.

Many properties of the zebrafish, including small size, large number of embryos, speed of development, and transparency of embryos, are favourable for genetic analysis. In addition, the ability to generate haploids with almost normal embryonic development allows the detection of recessive, early acting mutations without breeding homozygotes. Taken together, these properties should facilitate saturation screening for developmental mutants in the zebrafish. Until recently, such screens have not been performed, and as a result few zebrafish mutations have been described.

Despite this problem, the information obtained from the characterization of the *cyclops (cyc-1)* and *spadetail (spt-1)* mutations has clearly shown that genetic analysis of zebrafish is a fruitful approach for investigating vertebrate development.

One of these mutations, *spt-1,* which acts autonomously in prospective mesodermal cells to misdirect their gastrulation movements, hits also been used to investigate the regulation of homeobox genes. Despite the severe somitic abnormalities in the trunk region of *spt-1* embryos, staining with the XlHbox1 antibody indicated that the *hox-C6* gene was expressed within the appropriate AP domains in both the CNS and the mesoderm. The positional information responsible for regulating *hox-C6* seemed unaffected by the abnormal somitic segmentation caused by the mutation. In contrast, the expression of *eng* genes in muscle pioneers was affected in spt-1 embryos.

In the trunk region of mutant embryos, the myotomes were poorly patterned and lacked horizontal myosepta, as well as eng-expressing muscle pioneers. This observation supported the interpretation that cells expressing *eng* in the somite play an important role in the dorsoventral subdivision of the myotome. To elucidate the functions of the various homeobox and *pax* genes identified in zebrafish, it is necessary to apply transgenic methods. Transgenic zebrafish have been reported for several types of promoter-reporter constructions, further technical advances are clearly required before such methods can be efficiently exploited for functional studies of the developmental genes. Using *lacZ* as a reporter gene, various promoter-lacZ fusions have been used for transgenic analysis of the cis-acting elements involved in regulating the embryonic expression of the murine *Hox* and *En genes*. Permanent germ-line transformants of promoter-lacZ constructs have not been reported for any of the zebrafish homeobox or pax genes, but mammalian Hox promoters have been successfully analysed in mosaically transgenic zebrafish. Using the lacZ reporter gene, the activity of any promoter can in principle be assayed directly in microinjected embryos. Using this approach, Westerfield and colleagues have shown that the promoters of the murine *Hox-A7 (Hox-1. 1)* and human *HOX-C6* can specify expression within the same spatial domains in the CNS and paraxial mesoderm of zebrafish embryos as in mice, suggesting conservation of the promoter functions. In these experiments, the expression domains were defined on the basis of the mosaic expression observed in many embryos.

Chapter 7

Biological Membranes

In multicellular organisms, specialized cells elaborate and exploit particular components of basic membrane transport processes—as, for example, in nerve cells, where the regulated flow of ions in and out of the cell is used to generate and convey electrical signals.

The selectively permeable membranes surrounding organelles divide the interior of a eukaryotic cell into several metabolically 'sealed' compartments in which different processes can be isolated.

Over the past decade or so the role of the endoplasmic reticulum and Golgi apparatus in membrane synthesis, the direction of proteins to their correct location in the cell, secretion and intracellular transport has become much clearer. Energy generation is the function of mitochondria and chloroplasts and their membranes play a crucial part in their role as powerhouses of the cell. Transport directly across biological membranes is generally restricted to small molecules. Particulate matter, macromolecules such as proteins and large carbohydrates, and water and solutes in quantity can however, enter cells by the bulk transport process of endocytosis. In this process they are engulfed and taken into the cell in a membranous vacuole or phagosome derived from the plasma membrane, which effectively isolates them from the cytoplasm and prevents the entry of large amounts of foreign matter from upsetting the cell's delicately balanced internal environment.

The plasma membrane of animal cells provides anchorage points for the cell's internal cytoskeleton. The nuclear membrane is believed to provide sites at which the dispersed mass of chromosomal material can be attached in an orderly way as it is being transcribed and replicated between cell divisions. As well as its role in regulating the selective entry and exit of material, the plasma membrane is also the distinctive face a cell presents to other cells. It bears the molecules by which one cell recognizes another and is the point at which many chemical signals such as hormones, neurotransmitters and growth factors are first intercepted. Defects or alterations in receptors or in the subsequent links in the chain can have serious consequences. Such disturbances have been implicated, for example, in the development of cancer, in some forms of diabetes, and in atherosclerosis. On the other hand, many useful drugs, from muscle relaxants to tranquillizers, act either by mimicking or blocking the action of the body's natural chemical signal molecules at particular receptors. Although the physiological role of the cell membrane in the selective uptake of ions and nutrients has long been recognised, a full explanation of these properties in terms of its structure has only been achieved within the past 20 years. One obstacle to investigating the relationship between membrane structure and function is that biological membranes display their characteristic properties fully only while they remain intact and *in situ*. Many technical approaches have been brought to bear on the problem—electron microscopy; novel biochemical techniques for isolating membrane proteins; immunological and other methods of identifying membrane components in situ by labelling them with specific antibodies and other proteins specific for cell-surface components; observations of the physical and biological properties of isolated membranes and artificial membranes; spectroscopic analysis of the electron transport chains in mitochondrial and chloroplast membranes; sophisticated electrical recording techniques to measure ion flow across membranes; and most recently, the use of recombinant DNA technology for isolating the genes for membrane proteins so that they can be produced in quantity for molecular studies. Advances in understanding membrane structure and function have been essential to the progress of cell biology towards the point at which knowledge about cell structure and biochemistry gained by many different approaches can be integrated to provide an all-round picture of cellular organization and function.

The chief membrane lipids are phospholipids—relatively small molecules consisting of fatty acids linked to glycerol or sphingosine, which in turn are linked through a phosphate group to a headgroup such as serine, choline, ethanolamine, or the sugar alcohol inositol. The fatty acid portions, which consist of long hydrocarbon chains, form hydrophobic (water-repelling, 'insoluble', 'tails'; the serine, choline, etc. headgroups are polar and hydrophilic (water-attracting, 'soluble'). Lipids with this type of structure naturally orient themselves in an aqueous environment as a continuous closed bilayer with the hydrophobic groups to the centre. The lipid bilayer is the basis of all biological membranes. In itself it forms an impermeable barrier to many of the molecules and ions that cells need. The centre of the bilayer is intensely hydrophobic, forming a barrier to the passage of ions and most water-soluble molecules.

The bilayer is permeable to lipid-soluble molecules and to small uncharged molecules such as oxygen and nitrogen and to some other small molecules such as carbon dioxide and urea. It is impassable by most water soluble organic molecules, which include the essential nutrients a cell needs, and to ions of any sort. Somewhat surprisingly water itself seems to pass through the lipid bilayer fairly freely. The lipid bilayer of biological membranes is, however, provided with many molecular channels, gates and pumps, formed by some of the protein components of the membrane, through which water-soluble molecules and ions can be selectively allowed to enter and leave the cell. The physical properties of the lipid bilayer have important consequences for biological membranes. Lipid bilayers tend to close in on themselves to form a closed compartment. The membranes of living cells never have 'ends' but are continuous sheets delimiting the various internal compartments of a cell. The only way a biological membrane can grow, therefore, is by the insertion of new material into a preexisting membrane. The attraction of lipids for each other also means that membranes tend to be self-sealing if disturbed .

Early models of membrane structure positioned the protein in ordered layers on either face, but this type of structure was not particularly successful in explaining the biological behaviour of membranes. The generally accepted view of the organization of a biological membrane is called the fluid mosaic model. It describes a lipid bilayer in which float proteins, some spanning the width of the membrane, some attached only to the interior or exterior face. This type of organization provides a sound structural basis for known transport properties of the membrane: the proteins spanning the membrane provide hydrophilic sites isolated from the hydrophobic core of the lipid bilayer through which ions and small organic molecules can either passively diffuse or be actively transported.

The fluid mosaic model of membrane structure proposed in the early 1970s is now supported by a large body of experimental evidence and its predicted properties and behaviour accommodate observed biological phenomena very satisfactorily.

The-fluid part of the description refers to the fact that the membrane is not a rigid, static structure but a flexible assembly held together by many non-covalent interactions between lipid and lipid and protein and lipid. Both proteins and lipids can move 'sideways' in the plane of the membrane.

In general, lipids move freely, as do many proteins. The observed movement of proteins on the surface of cells was an important stimulus to the development of the fluid mosaic model.

Movement is, however, limited to the plane of the membrane. Once proteins have become inserted in a membrane they cannot change their orientation, and the lipids of one half (or leaflet) of the bilayer do not exchange with the other, except by a very slow flip-over of individual molecules. Biological membranes therefore possess 'sidedness', a physical and biochemical difference between their two faces that is determined at the time of their formation and strictly maintained thereafter. This asymmetry is of crucial importance, it defines the inside of a cell or organelle with respect to the outside and ensures, for example, that transport processes go only in the required direction.

The lipid composition of membranes is fairly similar from cell to cell although the proportion of the different lipids in different membranes varies. The type and number of membrane proteins, however, differ considerably between cell types, and between the various membranes within the cell. The protein component of the membrane determines the distinctive biological properties of the membrane and, ultimately, the behaviour of the cell or organelle it encloses.

Proteins that are firmly embedded in the membrane are called integral membrane proteins. They can only be removed by disrupting the membrane with detergent or organic solvents. The technique of SDS-gel electrophoresis—the electrophoretic separation of membrane proteins in the presence of the detergent sodium dodecyl sulphate—has now made it possible to isolate integral membrane proteins. Some integral membrane proteins span the membrane entirely (trans-membrane proteins), some are believed to be embedded mostly in either the cytoplasmic or external lipid layer. Membrane proteins involved in the passage of ions are generally known as ion channels and ion pumps, those involved in transporting sugars, amino acids and other metabolites to which the membrane is not intrinsically permeable are usually called transport proteins. The plasma membrane in particular, as the external face of the cell, contains receptor proteins for (depending on the type of cell) hormones and growth factors, antigens, neurotransmitters etc., 'recognition' proteins such as the histocompatibility antigens, and others involved in cell movement, cell adhesion, and many more specialized functions.

The membranes of mitochondria and chloroplasts are intimately involved in energy generation. The mitochondrial inner membrane contains the biochemical machinery for converting the energy obtained from the controlled breakdown of glucose during cell respiration into the universal chemical energy currency of the cell-adenosine triphosphate (ATP). Chloroplasts and the cell membranes of photosynthetic bacteria contain elaborate protein complexes for harvesting the light energy trapped by photosensitive pigments (chlorophyll in green plants, bacteriochlorophyll in bacteria) and converting it into stored energy in the form of ATP or using it to drive other processes.

Because it is at present impossible to obtain high-resolution atomic structures of membrane proteins in situ, the three-dimensional arrangement of an integral protein in a membrane is at present largely deduced from its amino acid sequence, supplemented by a knowledge of its physiological function, and, where possible, by direct visual evidence gleaned from electron micrographs using such modern methods as cryo-electron microscopy (at very low temperatures).

In regularly patterned structures, such as the outer coats of some viruses or the near-crystalline purple membrane of *Halobacterium*, the shape of the protein molecules and their arrangement in relation to each other can be retrieved in some detail from electron micrographs by modem image enhancement techniques. The amino acid sequence of a membrane protein can be interpreted in the light of structural features common to all membrane proteins. They all have certain peculiarities of structure that allow them to remain firmly anchored in the hydrophobic environment of the membrane—namely, one or more hydrophobic regions that are presumed to be the parts embedded in the membrane. Some proteins are anchored in the membrane by covalent bonds to membrane lipids. Many membrane proteins are glycoproteins, and the hydrophilic carbohydrate side chains invariably are found on the portion or portions of the protein protruding from the outer face of the membrane. This is now known to be the inevitable consequence of the way proteins are originally inserted into membranes at the time of their synthesis.

Many phospholipids and glycolipids spontaneously form a bilayer in an aqueous solution. In certain conditions, this self-seals to form liposomes—small hollow spheres of lipid 'membrane' that have proved of use as artificial model membranes in experimental studies.

As well as their usefulness in membrane research liposomes have also been tested medically as a way of delivering therapeutic drugs more efficiently to target cells. Non-lipid soluble material can be packaged into liposomes during their formation. Liposomes coalesce with cells' plasma membranes, delivering their cargo directly into the interior of the cell. Cells need to take up from their immediate surroundings small organic molecules such as sugars and amino acids for food and fuel and to get rid of waste products. They must keep their internal pH (acidity/alkalinity), osmotic environment and ionic balance (e.g. of the ions Na^+, K^+, Ca^{2+}, Cl^-, etc.) within the limits required for maintaining the integrity and function of the delicate biochemical machinery on which the life of a cell depends. All these processes are regulated by the plasma membrane.

The lipid bilayer of a biological membrane is intrinsically freely permeable only to lipid-soluble molecules, O_2 and N_2, and small uncharged polar molecules such as carbon dioxide, urea and glycerol. The lipid bilayer is, however, provided with many molecular channels, gates and pumps, formed by some of the transmembrane protein components of the membrane, through which water-soluble molecules and ions can be selectively transported in and out of the cell.

There are a variety of transport mechanisms in biological membranes. In general, transport is of two broad types. In passive transport or facilitated diffusion a solute or ion is conveyed across the membrane without the expenditure of metabolic energy. Passage of ions through ion channels or the uptake of nutrients by binding to certain carrier proteins is the most common form of passive transport. One example is the transport of glucose out of the side and basal membranes of epithelial cells lining the gut and into tissue fluids, mediated by a glucose carrier protein. Uptake of glucose from the gut itself, however, is an active process. Passive transport can convey material only down a concentration gradient (or for molecules carrying an electrical charge, an electrochemical gradient). To transport ions and solutes across the membrane against their concentration or electrochemical gradients, active transport processes are needed.

The internal and external cavities of the body are lined with epithelia (sing. epithelium). These are sheets of cells that effectively seal off the internal environment of tissues and organs from the exterior.

Epithelial cells are joined together by tight junctions that form a complete seal, preventing ions or molecules, however small, penetrating between the cells. Anything entering or leaving the body therefore has to be transported across the membranes of the epithelial cell, which form a barrier that serves much the same purpose in the body as a whole as the plasma membrane does in the cell. The way in which different transport processes localized to different regions of the cell membrane enable a cell to carry out a physiological task is illustrated by the epithelium lining the gut. Gut epithelial cells have a highly asymmetrical *(polarized)* architecture that reflects their function. They take up material from the apical face and deliver it into the circulation-via their basal and lateral faces. The different faces of the epithelial cell represent different membrane domains to which particular transport proteins are restricted. The apical face is separated from the rest of the membrane by the tight junctions. Here the membranes of two adjoining cells are so closely apposed that the seal they form is a barrier to diffusion of proteins from apical to lateral and basal faces. Nutrients such as glucose and amino acids—the products of digestion in the lumen of the gut—are taken up against a concentration gradient by active transport through the apical membrane of the epithelial cell, which faces into the gut and is special zed for this purpose with a highly infolded surface offering a large area for absorption. Uptake of glucose and amino acids across the apical membranes is mediated by sodium symports. The carrier protein for glucose in the apical membrane has two separate binding sites, one for Na^+ and the other for glucose. Sodium tends to enter the cell down its concentration gradient and as it does so drags glucose in with it.

Ions play an important part in the life of a cell in their various roles as cofactors for metabolic processes, determinants of the cell's internal osmotic environment, and as carriers of electrical charge. Hydrogen ion (proton, H^+) concentration determines the cell's internal pH, and its flow across certain membranes also drives the synthesis of ATP.

Maintaining an internal environment of suitable ionic composition, pH and electrical neutrality overall means a continual flow of ions in and out of the cell, as the concentration of a particular ion inside a cell is usually very different from its concentration in the surrounding medium. The ions concerned are chiefly the cations sodium (Na^+), potassium (K^+) and calcium (Ca^{2+}) and the anions chloride (Cl^-) and bicarbonate (HCO_3^-). Magnesium (Mg^{2+}) is an important cofactor of many metabolic reactions. Ion movement in and out of a cell is controlled through the plasma membrane, which contains several types of ion pumps and ion channels—protein assemblies each specific for a particular ion (or ions) and that regulate an ion's entry and exit in a variety of different ways. The numbers and types of pumps or channels vary from cell to cell, depending on its particular specialization. Electrically active cells, such as nerve and muscle, that can propagate an electrical impulse, do so by virtue of the properties of the various ion channels studding their membranes. The concentration of a particular ion inside a cell is usually different from its concentration in the surrounding medium, whether it is the intercellular fluid of the animal body, or the watery environment of free-living microorganisms. Animal cells maintain a K^+ content of up to 40 times greater and a Na^+ content of up to 15 times lower than that of the fluid that bathes them. They need the potassium for essential metabolic processes; sodium, on the other hand, inhibits many of these processes. The inside of a cell must be electrically neutral. Therefore many cells also maintain a Cl^- content much lower than the external medium, because they have to get rid of excess negative electrical charge—many of their fixed constituents that are too large to leave the cell (e.g. nucleic acids, proteins, and especially the phosphate groups—and carboxyl groups they carry) are negatively charged. Cells also maintain a very low internal concentration of free Ca^{2+} ions compared with the external environment. The inevitable diffusion of ions down their concentration gradient (i.e. from a region of high to a region of low concentration) and the 'leakiness' of the cell membrane to ions results, for example, in a continual slow leakage of K^+ out of and Na^+ into the cell. The cell maintains the ionic balance it requires by actively pumping ions in and out against their respective concentration gradients.

Ion pumps are specific for particular ions, and the best characterized is the sodium-potassium pump or sodium pump of ~the animal cell plasma membrane, also known as Na^+ - K^+-ATPase. This simultaneously moves sodium out of and potassium into the cell against their concentration gradients and in doing so helps to generate the membrane potential. The sodium pump is believed to consist of a pair of identical catalytic subunits and a pair of transmembrane glycoproteins of unknown function.

The ATPase activity is located on the cytoplasmic face of the catalytic subunit. For every molecule of ATP hydrolysed, three Na^+ are transported out of the cell and two K^+ are transported inwards. Binding of Na^+ and ATP hydrolysis is believed to result in phosphorylation of the catalytic subunit and a subsequent change in its conformation which releases the Na^+ to the outside. Binding of external K^+ leads to dephosphorylation and the protein returns to its original state, releasing K^+ to the cytoplasmic side of the membrane. A typical resting animal cell devotes a third of the total energy it produces to fuelling its sodium-potassium pumps, a measure of their importance to the cellular economy. Calcium pumps are also important in cellular physiology. The ion gradients that cells maintain across their membranes result in a difference in electrical potential across the plasma membrane (the membrane potential), the interior of the cell being electrically negative with respect to the outside. Membrane potential is conventionally measured by inserting a fine glass capillary microelectrode into the cell and another electrode into the extracellular medium. The potential difference between the two is amplified and fed into an oscilloscope. It ranges between -20 and -200 mV, depending on the type of cell. The membrane potential in most animal cells is generated and maintained largely through the activity of the sodium-potassium pump (Na^+ ,K^+-ATPase).

Cells tend to lose K^+ and gain Na^+ and to maintain the correct balance the Na^+, K^+-ATPase in the plasma membrane is continually actively pumping Na^+ out and K^+ in against their concentration gradients. But K^+ is continually leaking out of the cell down its steep concentration gradient through a permanently open K^+ ion channel (the K^+ leak channel). This loss of K' (positive charge) leads to the interior of the cell becoming electrically negative with respect to the exterior—i.e. a membrane potential is generated. This then tends to retard the loss of K^+ from the cell—as K^+ is positive it will not move passively against the electrical gradient—and, instead, K^+ tends to enter the cell. At a particular membrane potential, the tendency of K^+ to move out of the cell against its concentration gradient is exactly balanced by its tendency to enter the cell down its electrical gradient, and the membrane potential stabilizes at this value called the resting membrane potential.

However, the K^+ channel is also slightly permeable to Na^- and so there is also a continual slow flow of Na ' into the cell. If it were not for the Na^+ - K^+ pump actively maintaining the concentration gradients across the membrane, this would eventually lead to the equilibration of Na^+ and K^+ across the cell membrane and the disappearance of the membrane potential.

In natural circumstances small disturbances of this sort result from the actions of many extracellular signals, which lead to ion flows across the plasma membrane that are often involved in triggering the cell's physiological response. In electrically active cells, however, a disturbance of membrane potential has dramatic results, leading to a sudden discharge—an electrical impulse. The membranes of all cells are studded with ion channels—aqueous pores traversing the membrane, isolated from the surrounding hydrophobic lipid bilayer by the protein walls of the channel protein. In contrast to ion pumps, movement of ions through ion channels requires no expenditure of metabolic energy. Ions flow through the pores by simple diffusion down their electrochemical gradients. Ion channels are selective for different ions and those that are permanently open (such as the K^+ leak channel) contribute to the inherent 'leakiness' of cell membranes. There is a basic distinction between those that allow cations (e.g. Na^+, K^+ and Ca^{2+}) to pass and those that allow anions (e.g. Cl^- and bicarbonate HCO_3^-).

The molecular structure of ion channels has only been determined in the past few years. The amino acid sequences of several channel proteins have now been determined indirectly from cloned cDNA. The existence of many other ion channels is inferred from detectable flows of electric current in or out of a cell (conductances). The sophisticated technique of patch-clamping, in which a minute patch of membrane containing a presumed single channel is electrically sealed off from, its surrounding membrane across the mouth of a hollow glass microelectrode, now allows recording of current flow through a single channel. Some channels are not permanently open but open transiently only in response to a particular stimulus. They figure largely in cells' responses to many external and intracellular stimuli, the resulting ion flows in and out of the cell triggering the biochemical machinery that produces the cell's typical response.

There are two functional classes of gated ion channel—those that open in response to binding of a signal molecule (ligand-gated channels) and those that open in response to a change in membrane potential (voltage-gated channels). They may stay open for times varying from a fraction of a millisecond to a few milliseconds. By analogy with the behaviour of many other proteins, it is believed that binding of another molecule (the ligand) or a voltage change in the surrounding membrane induces a conformational change in the three-dimensional structure of the channel protein that opens the aqueous pore through its centre. Selective opening of ion channels is a part of many cells' response to external stimuli, and gated ion channels probably exist in all cells. They have been best studied so far in nerve and muscle where electrical impulses are generated by the opening of voltage-gated channels, and several neurotransmitters act at ligand-gated channels. A special type of ion channel, which also allows other small molecules to pass, is the gap junction protein. The function of gap junctions in intercellular communication is described elsewhere. The ability of an ion channel to respond to its particular stimulus or to convey ions can be altered by biochemical modifications to the channel protein on its intracellular face. Such modifications, which can permanently or semi-permanently shut down a set of ion channels in the plasma membrane can be induced by various stimuli and produce long-term changes in a cell's properties and behaviour.

The best-studied ligand-gated channels are those that comprise the receptors for several neurotransmitters on the plasma membranes of nerve and muscle cells. The receptors for acetylcholine on skeletal muscle cells, and for the neurotransmitters GABA (-γ-aminobutyric acid) and glycine on neurones are ion channel proteins. The amino acid sequences of these ion channels have been resolved and show striking similarities. This suggests that they, and possibly other neurotransmitter-activated ion channels whose sequences are not yet known, belong to a 'superfamily' of proteins descended from a common ancestor. During evolution they would have diversified to come under the control of different compounds and to convey different ions. Other ligand-gated channels have been found that are responsive to Ca^{2+} and cyclic nucleotides.

Voltage-gated channels open in response to a change in the membrane potential in their vicinity, opening only when the membrane potential reaches a certain 'threshold' value. Part of the channel protein is involved in 'sensing' voltage changes, presumably by a change in conformation. The hest-studied voltage-gated channels are those from the membranes of electrically excitable cells—nerve and muscle—where they are essential for generating and propagating electrical impulses, and in regulating the release of neurotransmitter from neurone terminals. The structure of the main voltage-gated Na^+ channel from the electric organ of the electric eel *Electrophorus* (a favourite material of neurophysiologists) has recently been deduced from its amino acid sequence. Many drugs and poisons are known to exert their effects by physically blocking ion channels and preventing ion flow. Local anaesthetics and 'ganglion blockers' used as anaesthetics inhibit the conduction and transmission of signals in nerve cells, and anti-dysrhythmic drugs used to treat certain heart conditions block certain ion channels in heart muscle. As well as transporting ions and small molecules directly across the plasma membrane via specific membrane channels, pumps and carrier proteins, cells can also take in bulk quantities of the extracellular medium, macromolecules and, even large particles such as other cells, by the quite separate process of endocytosis. Some cells are specialized for phagocytosis.

Some free-living protozoins (such as the familiar *Amoeba)* take in food in this way; in multicellular animals, specialized phagocytes (such as macrophages and polymorpho-nuclear leukocytes in mammals) scavenge dead cells and cell debris, and are also a first line of defence against invading microorganisms. Pinocytosis, on the other hand, is a routine occurrence in virtually all animal cells—to the extent that a macrophage internalizes the equivalent of the whole of its plasma membrane in endocytotic vesicles every half hour or so, non-phagocytic cells being somewhat less active.

Endocytosis is used by mammalian cells in various ways. Soluble material from blood, for example, is bulk transported across the endothelial cells of the walls of small blood vessels into the tissue fluid on the other side by endocytosis at one face of the cell, transport across the cell in endocytotic vesicles (transcytosis), and delivery into the extracellular fluid at the other face. Sometimes the endocytotic vesicles fuse to form a continuous channel through the cell.

As well as bulk transport of solute, certain molecules can be taken up by the selective process of receptor-mediated endocytosis. Molecules binding to a specific receptor protein at the cell surface trigger its internalization as part of the membrane of an endocytotic vesicle. Receptor-mediated endocytosis occurs in many mammalian cells. It is, for example, the means by which they take up cholesterol. All animal cells need cholesterol as a constituent of membrane and in mammals it is manufactured in the liver and transported via the bloodstream in the form of so-called low-density lipoprotein (LDL) particles. These contain a core of cholesterol molecules esterified to fatty acids and surrounded by a lipid bilayer into which is inserted the protein component (apolipoprotein) of the particle. Cells requiring cholesterol to manufacture new membrane display special LDL receptor proteins on their surface. When these bind the LDL protein the whole complex is then internalized and enzymes in the endocytotic vesicle release the cholesterol for the cell's use. The receptors and other membrane components are then recycled, it is presumed back to the plasma membrane or via the Golgi bodies.

Cholesterol accumulating in the blood contributes to the formation of atherosclerotic plaques on the walls of arteries. Buildup of atherosclerotic plaque narrows the artery, impeding blood flow, raising blood pressure and so predisposing to thrombosis. When it occurs in coronary arteries, it can lead to heart failure. Various genetically determined defects in the receptor mechanism for cholesterol uptake and a consequent high level of cholesterol in the blood are now known to be the cause of the inherited predisposition to severe premature atherosclerosis, called familial atherosclerosis. Those suffering from this condition often die at an early age from coronary artery disease. Many of the receptors that cells possess to intercept chemical signals such as hormones at the cell surface are also 'turned-over' by receptor-mediated endocytosis once a signalling molecule has bound. Most of these signalling molecules are in fact believed to act at the cell surface and the signal transmitted rapidly via the receptor to intracellular biochemical pathways. The relatively slow process of receptor-mediated endocytosis is therefore believed to represent an internalization of used-up receptors for recycling and is also an important way in which cells can control the number of receptors on their surface and thus their sensitivity to a particular chemical message.

Receptor-mediated endocytosis does not apparently occur at random on the cell surface but is confined to areas called coated pits, with which the receptors are associated or to which they migrate after binding their ligand. Coated pits endocytose to form coated vesicles. The 'coat' of coated vesicles and pits is a polyhedral lattice formed mainly of the protein clathrin. Coated vesicles are also involved in vesicular transport between endoplasmic reticulum and the Golgi bodies and the Golgi bodies and the plasma membrane. The continual process of endocytosis is important in animal cells as a means of recovering 'excess' membrane added to the plasma membrane during secretion. Plant cells, however, do not routinely use endocytosis as a means of taking up material from outside the cell. (It would have little useful purpose as the cell wall prevents particles and many large molecules from reaching the plasma membrane.

Endocytosis is in any case energetically unfavourable in plant cells where the plasma membrane is pressed tight against the cell wall as a result of the high internal osmotic pressure (turgor pressure) developed in the cell due to the presence of a restraining rigid cell wall.) Nevertheless, all plant cells add large amounts of membrane to the plasma membrane during secretion of cell wall material, and it is believed that, by analógy with animal cells, this excess membrane is recovered by endocytosis of coated pits that have been seen in plant cell plasma membrane preparations.

Most cells manufacture and release material to the exterior—the process generally called secretion. Eukaryotic cells variously secrete digestive enzymes, protein, peptide and steroid hormones, polysaccharides (for plant cell walls), the materials of the extracellular matrix, the proteoglycans of cartilage, and a wide variety of different substances used ás neurotransmitters in the nervous system. Depending on the type of cell these vesicles are called secretory vesicles, secretory granules, zymogen granules (which contain inactive digestive enzyme precursors) or synaptic vesicles. The intracellular route by which material is packaged into secretory vesicles depends on the type of material. Proteins, polysaccharides and proteoglycans are packaged via the endoplasmic reticulum and Golgi bodies.

Small molecules such as the catecholamine and amino acid neurotransmitters, on the other hand, are synthesized in the cytoplasm and actively transported into preformed vesicles. In many some cases, secretion is continuous—the manufacture and release of antibodies by the plasma cells of the immune system is a case in point. In many other cases, however, material is synthesized and stored in secretory vesicles and only released when the cell is stimulated by an appropriate signal, such as a hormone or nerve impulse. The chain of intracellular events linking hormonal or nervous stimulation and exocytosis is at present receiving much attention from biochemists and cell biologists. A common factor triggering exocytosis in many cells in response to a variety of stimuli is a sudden rise in the concentration of Intracellular free calcium ions. The means by which Ca^{2+} triggers exocytosis is still unclear. It probably induces the fusion of transport vesicle and plasma membrane indirectly through its interaction with ubiquitous calcium-binding proteins such as calmodulin, and/or by its actions on components of the cytoskeleton that are believed to play a role in conveying secretory vesicles to the plasma membrane.

Chapter 8

Petal and Stamen Development

Petals often are the most conspicuous organs of the flower. In many species, the large, brightly coloured petals are thought to have evolved to maximize visual and olfactory signals to attract pollinators. Goethe, was one of the first to articulate the hypothesis, proposing that plants are composed of a series of similar organs that undergo a progressive metamorphosis. Goethe theorised that this metamorphosis was brought about by the action of a sap that became refined as it progressed up the plant culminating in the successive development of the different types of floral organs.

More recent phylogenetic analyses have refined this view and suggested that while carpels and stamens appear to have each evolved prior to the origin of the angiosperms, petals and sepals may have been derived more recently, and in particular, petals are thought to have been derived from other floral organs.

In the majority of angiosperm groups, including all of the higher eudicots and monocots, petals are thought to be derived from stamens, a condition termed andropetaly. Among the lower eudicot and magnolid dicot families, there are species that contain andropetals, as well as species in which petals appear to have been derived from sepals or other leaflike organs, termed bracteopetaly.

The predominance of andropetaly in the angiosperms suggests that, in the majority of angiosperm species, the developmental programs controlling the specification of petal and stamen identities are similar and evolutionarily linked. Almost all of the experimental work on flower development that has been carried out has involved higher eudicot species, including *Arabidopsis, Antirrhinum* (snapdragon), Nicotiana (tobacco), and *Petunia.*

The emerging genetic and molecular evidence indicates that there are several genes that are uniquely required for the development of both petals and stamens in these species, supporting the idea that a common molecular mechanism is involved in the specification of these two organ types.

The differentiation of particular cell types in the flower also depends on intercellular interactions. Because the ultimate fate of a plant cell depends on its final position and not its other to coordinate their development and differentiation.

In addition, cell fates can be respecified until their final division, indicating that intercellular interactions must be occurring throughout development. Investigation of flower development, in particular, have been instrumental in beginning to elucidate some of the underlying molecular mechanisms by which plant cells signal to each other. Several lines of evidence indicate that such intercellular interactions are critical for petal and stamen development and will be reviewed herein.

Petals and stamens arise from cells derived from the florally determined shoot apical meristem. The floral meristem of dicots and some monocots is organized into three cell layers, which generate the tissues of the flower. The L1, or outer, layer of meristematic cells gives rise to the epiderms, whereas the underlying L2 and L3 cells contribute to the niesophyll and vascular tissues of the flower. Cells in all three cell layers of the flanks of the floral ineristem change their patterns of growth and division to form the floral organ primordia.

In general, the floral organs are initiated sequentially with sepals arising first followed by the almost stimutaneous appearance of petal and stamen primordia, and finally the carpel primordia become apparent. Although all three cell layers contribute to the development of' the petals and stamens, the contribution of each cell layer can be somewhat variable. Marked cell lineages have been used to show that each cell layer gives rise to different extents of an organ. Furthermore a cell in one invade another layer and differentiate according to its new position, and such invasion events appear to occur stochastically.

These observations raise several questions about how floral organ primordia are specified. Because subsets of meristematic cell, proliferate in specific regions on the flanks of the meristem, some mechanism must be acting to define regions of primordial outgrowth. Furthermore, these regions are bounded by cells that apparently do not proliferate to the same extent.

Within a given species, the number of floral organ primordia is almost always constant, indicating that the processes defining organ primordia must be tightly regulated. Despite this orderly pattern of growth, the occurrence of cell layer invasion events indicates that floral primordial cells are not committed to a particular developmental fate. Several types of such spatially regulated signaling events have been proposed to explain the regular pattern of organ primordium initiation, as well as how the development of cells within a floral organ primordium is coordinated. These include signals between whorls of organ primordia, signaling between organ primordia within a whorl, and signaling between the cell layers of a developing organ. In addition to spatial control of signaling, it appears that such inductive interactions may be regulated temporally as well.

The result of experiments in which meristems have been manipulated surgically have been interpreted to suggest that signals between whorls are necessary for the development of the floral organ primordia. Researchers have suggested that inductive signals from one whorl tire required to promote the formation of the next whorl of organ types. For instance the incipient sepal primordia were Postulated to induce the formation of the second whorl petal primordia. These models, however, did not clearly distinguish between the specification of an organ primordium in a distinct position and the specification of organ identity.

It is clear from the existence of floral homeotic mutants, which can have organs of tiny identity in ally position, that such inductive interactions are not required to specify particular organ identities. These types of interwhorl inductive signals may, however, operate to specify the position of the next whorl of organs. Furthermore, these models do not attempt to explain how individual organs arise in specific locations on the flanks of the meristem, not have any corresponding signalling molecules been identified. Genetic ablation studies indicate, however that if such interwhorl signals exist, they can only be operating at very early times in floral development. Signaling between organs within a whorl account for how the spacing of organ primordia in a particular whorl occurs. Evidence supporting such intrawhorl signalling comes from experiments in which developing tobacco floral meristems were bisected. Depending of the time of surgery, half-meristems could regenerate some or till of the missing organs, implying that the pre-existing organ primordia of the unoperated side of the half-apex were sufficient to induce the development of' new organs on the operated side. Again, the nature of such an interwhorl signal has not yet been established. Both the interwhorl and intrawhorl signals postulated to specify the inception of organ primordia in particular locations Could be similar or identical to those signals that have been proposed to operate in the positioning of leaf primordia during vegetative development. Phase-specific cues must be overlaid upon such a positional signaling mechanism in order to specify leaf or floral organ primordia.

Within a floral organ, it does appear that intercellular signals operate to coordinate development. Periclinal chimeras, in which one cell layer is genotypically different from the other cell layers, have been used to demonstrate that interlayer signaling, does occur within developing floral organ primordia. Camellia periclinal chimeras in which the L1 was derived from a species containing stamens and carpels, whereas the L2 and L3 were derived from a sterile line lacking stamens and carpels, develop normal stamens and carpels. In tomato, meristem size depends, to a large extent, on the genotype of the 1-3 layer. Specific genes also have been shown to play a role if, such interlayer signaling. For instance, periclinal chimeras in which the cell layers differ only by the function of' a single gene have been used to show that the product of the *Antirrhinum FLO* gene acts non-cell-autonomously, and its presence in just the LI layer of the flower can almost completely rescue the phenotype of otherwise genetically flo mutant floral meristem cells. These results indicate that there are likely to be inductive interactions between all layers of cells in the floral meristem, although the mechanisms by which such cues are disseminated are far from clear. It also is not clear whether there is a temporal window during which such interlayer interactions have to occur. What all of these studies point out is that the development of the floral organs appears to rely on a complex interplay between different cells. Inductive interactions may occur to specify the correct positioning of organ primordia, whereas within an organ intercellular interactions must be acting to coordinate the development of different tissue types.

Ongoing molecular and genetic analyses are beginning to identify many of the genes required for various aspects of floral development. It seems clear that at least a subset of these genes will be involved in producing or responding to those signaling molecules that are required for organogenesis.

The next sections will review the molecular and genetic analyses that have been carried out to identify genes involved in elaborating, petal and stamen organ identities. In a few cases, the products of several of these genes have been implicated in the cell-cell signaling events required to coordinate the division and differentiation of cells in a developing organ primordium.

A number of mutations have been identified that cause homeotic transformations of the floral organs. In *Arabidopsis,* the *APETALA3 (AP3)* and *PISTLLATA (PI)* genes are required to specify petals and stamens. Mutations in either of these genes result in a homeotic conversion of petals into sepal-like structures, and stamens are converted into carpels. The homologous genes in *Antirrhinum, DEFICIENS (DEF)* and *GLOBOSA* (GLO)), not only are similar to their counterparts at the sequence level, but also are required to specify petal and stamen identities in this species. Homologues of *AP3* and PI also have been identified in a number of other angiosperms, and at least within the higher eudicots, it appears that the functions of these genes in specifying petal and stamen identities are conserved. On the basis of the phenotypes produced by several floral homeotic mutations, a combinatorial model (the ABC model) has been proposed to explain how floral or-an identities are established. Briefly, combinations of homeotic gene function, in different regions of the developing flower are thought to act to specify particular of an identities. These functions are proposed to correspond to the actions of particular sets of genes.

In *Arabidopsis,* the two A group genes, *APETALAI (API)* and *APETALA2 (AP2),* are both required for sepal development. AP3 and *Pl,* the B group genes, in combination with the A group genes act to specify petal identity, whereas the B group genes, in combination with the action of the C group gene, specify stamens. Lastly, the C group gene, *AGAMOUS (AG),* is required in the fourth whorl to specify carpel identity. In addition, the A and C functions are proposed to negatively regulate each other. The phenotypes produced by mutations in these floral homeotic genes, or in doubly mutant combinations, are largely consistent with the ABC model. The *AP3, Pl, AG,* and *API* genes have been cloned, and they share homology with a family of yeast and animal genes that encode DNA binding transcription factors. The 56-amino acid region of homology has been termed the MADS box domain and corresponds to the DNA binding domains of the yeast and mammalian gene products MADS box domain containing transcription factors binds to a Consensus DNA sequence, $CC(A/T)_6GG$, which is called the CArG box. Binding sites for several of the plant MADS box containing gene products have been defined *in vitro,* and these fit the known CArG box consensus motif. These observations suggest that the plant MADS box gene products may act as transcriptional regulators. Among the plant MADS box containing genes, a second region of homology has been identified and called the K box, which may be required for protein-protein interactions. The *AP2* gene also has been postulated to encode a transcription factor, because it contains two copies of a motif that is similar to the DNA binding domains of ethylene response element binding proteins. In *Arabidopsis,* then, *AP3* and PI are uniquely required for specifying petal and stamen identities and presumably act as transcription factors to regulate the expression of other petal- and stamen-specific genes. Although neither the AP3 nor PI protein call bind to DNA independently, AP3-PI heterodimers have been demonstrated to bind to CArG box sequences *in vitro.* Similarly, the products of the *Antirrhinum DEF* and GLO genes also bind to DNA as a heterodimer. Several other lines of evidence suggest that AP3 and PI must interact with each other to mediate their proposed functions. AP3 and PI both need to present together in the same cell for AP3 and PI fusion proteins to be localized to the nucleus. Epitope-tagged AP3 protein can co-immuno-precipitate PI, and conversely, epitope-tagged PI can co-immunoprecipitate AP3. Finally DEF and GLO can interact in the yeast two-hybrid system, further supporting the contention that these classes of proteins bind to each other to effect their function.

The transcriptional expression patterns of the B group genes generally are consistent with their roles in specifying petal and stamen identity. AP3 transcripts are seen early in floral development in the meristematic domain that is thought to give rise to the petals and stamens. AP3 expression persists in the developing petal and stamen primordia until the flowers mature, with transcript levels declining at the time of fertilization. PI transcripts are also detectable early in floral development, but are initially found in the meristematic cells that will give rise to petal, stamen, and carpel primordia. As the organ primordia emerge, this pattern of expression becomes refined, with

PI transcripts becoming limited to the developing petals and stamens. That the AP3 and PI expression patterns become coincident appears to reflect the fact that the AP3 And PI gene products are required for both auto and cross regulatory interactions.

A functional AP3 gene product is required to maintain both AP3 and PI transcription and similarly, PI is required to maintain AP3 and PI transcription. The coordinate expression of AP3 and PI together appears to be sufficient to confer petal or stamen identify in the flower, because ectopic expression of both AP3 and P1 in the first whorl is sufficient to confer petal identity. Ectopic expression of AP3 in the fourth whorl, where P1 is already expressed, results in the development of stamens. In the case of AP3, this maintenance of expression presumably is mediated in part by the binding of an AP3-PI heterodimeric transcription factor to at least three sequences in the AP3 promoter. In some other higher eudicot species, the expression patterns of the AP3 and P/ homologues are reversed. For instance, in Antirrhinum, DEF, the orthologue of AP3, is expressed initially in the petal, stamen, and carpel primordia, whereas expression of the PI orthologue *GLO* is restricted to just the developing petals and stamens. A similar switch also has been seen in the expression of B group genes in the Solanaceae. Despite this switch in expression patterns, the functions of these genes appear to be similar to those of their Arabidopsis counterparts. This presumably reflects the fact that the overlap in B group gene expression patterns, and consequently the expression domain that can be maintained by auto- and cross-regulatory mechanisms, is the critical factor in establishing petal and stamen identities. The restriction of AP3 gene expression to just floral tissues may also depend on epigenetic mechanisms.

Mutations in the *Arabidopsis CURLY LEAF (CLF)* gene result in ectopic AP3 expression in the leaves. *CLF* encodes a product with similarity to the Drosophila Polycomb group of proteins, which have been proposed to regulate gene expression though modulating chromatin structure.

Alterations in methylation status also have been implicated in AP3 regulation, because plants in which genomic cytosine methylation levels have been reduced display ectopic AP3 expression. In addition to maintaining B group gene expression within a developing organ primordium, some mechanism must ensure that the boundaries of B group gene expression are in register with whorl boundaries. It appears that these boundary conditions are established early in floral development. Lineage analyses in *Antirrhinum* have shown that a lineage restriction arises between whorls at the same stage that B group gene expression commences. This observation suggests that either cells are dividing more slowly or they have altered planes of division at the whorl boundary, thus restricting the proliferation of cells into the adjacent domain.

Although it is possible that the B group genes function to establish these boundaries, it seems unlikely because B group gene expression is just commencing at the time at which these lineage restrictions are being laid down. An alternative possibility is that another gene operates to set the whorl boundary, and this same gene also functions to delimit the boundary of B group gene expression. The product of the *Antirrhinum FIMBRIATA (FIM)* gene may be playing this role *F/M* is expressed at whorl boundaries and is required for B group gene expression. The product of the *FIM* gene has been implicated in the control of cell division, because it can complex with proteins that have homology to yeast gene products required for protein degradation and cell cycle progression The *Arabidopsis homologue of FIM,- UNUSUAL FLORAL ORGANS (UFO)* also has been postulated to regulate the domain of B group gene expression in a similar fashion. At least part of the repertoire of B group gene function may include the regulation of cell division within the B group domain. It is clear that some aspects of cell proliferation within an organ can be controlled by B group gene function, as over expression off *PMADSI*, the *AP3* homologue in *Petunia*, stimulates extra lateral cell divisions. Furthermore, *SUPERMAN (SUP)*, which encodes a zinc finger domain containing protein, appears to control the rates or planes of cell division at the third-fourth whorl boundary, as mutations in SUP result in excessive proliferation of third whorl cells. AP3 gene function is required to activate SUP expression, and, in turn, SUP activity itself appears to be required to restrict *AP3* expression to the third whorl. This feedback control mechanism to regulate the boundary of the third whorl may rely in part on intercellular interactions, because the patterns of SUP and B expression are not entirely coincident.

Coordination of the development of the different cell layers in the petals in stamens appears to depend on intercellular signaling mediated by B group gene activity. Periclinal chimeras have been used to assess whether expression of a particular floral homeotic gene in a single cell layer of the developing floral bud is sufficient to confer a wild-type phenotype. Expression of the Antirrhinum DEF and GLO B group genes in just the inner L2 and L3 layers is sufficient to induce the formation of petaloid epidermal cells in the L1 layer of the second whorl. The fact that the DEF and GLO proteins can be found in the L1 layers of these chimeric plants suggests that these floral homeotic proteins are being translocated between cell layers via the plasmodesmata. Despite the apparent ability of floral homeotic proteins to move between cell layers, there is no evidence for lateral intralayer signaling. Revertant wild-type *DEF* or GLO sectors in an otherwise mutant L1 layer are sharply bounded, indicating that intralayer movement either does not occur or is limited. The directional cell-to-cell communication between layers may reflect the distribution of plasmodesmata in the developing flower bud.

Within the floral meristem, cells are connected within a cell layer by primary plasmodesmata; however, cells in different layers must form secondary plasmodesmata to allow the interlayer passage of molecules. This difference in plasmodesmatal origin allows for the possibility of regulated trafficking in only certain directions within the developing floral meristem. The continued expression of the floral homeotic genes in particular organs presumably serves to direct the expression of all array of target genes. Few such candidate downstream target genes have been identified, however.

There are a number of problems associated with identifying such target genes. First, the floral homeotic genes are expressed throughout much of floral development and have been shown to have a functional role late in floral development as well. Because it is likely that the repertoire of target genes is changing over time, this implies that the specificity of floral homeotic gene function must also be modulated during development. A second difficulty lies in the fact that the DNA binding specificities of AP1, AG, and AP3-P1 dimers are very similar in vitro, making the identification of targets using in vitro binding approaches unreliable. In addition, expression of chimeric AP1, AP3, PI, or AG cDNAs containing heterologous MADS box domains does not affect their homeotic activity, indicating that DNA binding specificity alone is not a good predictor of in vivo function. The specificity of the floral homeotic *gene* products for particular downstream target sequences may that may modulate DNA binding activity in vivo. Precedent for this exists in yeast and mammalian systems, in which MADS domain containing gene products have modified DNA binding specificities depending on the presence of' accessory factors. In fact, most of the putative downstream targets of genes have been identified by other criteria and only later the floral homeotic, shown to be possible targets of the floral homeotic genes. However, a novel approach to identify genes specifically upregulated in response to homeotic gene action in the absence of protein synthesis has led to the identification of one putative direct target of the AP3 and P1 floral homeotic genes. Although a number of petal-and stamen-specific differentiation- associated genes have been described, much work needs to be done to demonstrate how these genes are regulated by the floral homeotic genes.

Petals are relatively simple laminar structures, with a sheet of epidermal cells covering an internal layer of mesophyll. Petals develop by divisions in all three meristematic cell layers, and the formation of the lamina depends on division in submaruinal initial cells. Cell divisions cease in the developing petal at a relatively early stage, and the final phase of petal growth depends entirely on cell expansion. The terminally differentiated petal mesophyll cells usually are uniform in size and generally rounded, whereas the petal epidermis lacks stomata and trichomes and is composed of regular cone-shaped epidermal cells. Generally, the epidermal cells are pigmented, and in some species there can be very elaborate patterns of pigmentation across each petal. For instance, many species have petals in which the base is differently pigmented from the tip of the petal. Because these coloration patterns are not clonal and pigment biosynthesis occurs after cell division has ceased, these patterns must reflect some underlying mechanism that functions to coordinate expression across a region of petal cells.

Despite the simple laminar structure, petal shape and size can vary dramatically in different species. Many of the elaborate petal shapes appear to have evolved to maximize pollinator attraction. Individual petals can be shaped differently within a flower, resulting in bilaterally symmetric flowers as in the pea family.

In many species, one or more petals are modified from a spur, which can be a very pronounced saclike structure. For instance, in some orchids, one petal forms a spur, which gives these flowers their characteristic appearance and serves to guide insects, bats, or other pollinators to the pollen. Some flowers have fused petals that form a tube, such as tobacco, whereas other species have what are termed "window" flowers.

Ceropegia species have elaborate window flowers in which the petals are fused at the tips, forming lateral openings, or windows, through which insect pollinators can reach the center of the flower. The fact that individual petals within a flower can have different shapes or intricate color patterns indicates that an underlying pattering, mechanism is operating to specify regional differences in those few cell types that make in the petal. A number of genes have been identified that are expressed predominantly or solely in the petals, and several mutations affecting different aspects of petal development also have been characterized. The analyses of the corresponding genes, as well as investigations is into differential gene regulation, have provided inroads into elucidating how regional differences in petal development are established.

Many flowers are zygomorphic, or bilaterally symmetric, and this departure from radial symmetry most often is exemplified in radical changes in the form and size of the petals. For instance, in *Antirrhinum,* the two dorsal petals shaped differently and have larger lobes than the lateral and ventral petals, contributing to the elaborate shape of these flowers. In addition to differences in shape between petals, the individual petals themselves can have an asymmetric shape. Flower symmetry has been shown to have a functional role as well, in that it can be a factor in the discrimination of pollen sources by insects.

The genetic control of zygomorphy has been investigated recently in *Antirrhinum,* although interest in understanding the genetic basis of zygomorphy dates back to Darwin. It appears that the asymmetry across the flower, its well as the asymmetric shape of individual petals in zygomorphic *Antirrhinum* flowers, is controlled by the same underlying genetic mechanism.

The Antirrhinum *cycloidea* (Q-c) mutation causes a partial transformation of dorsal petals into more lateral, and lateral petals into smaller, more ventral-like form. Each partially transformed petal itself also shows a graded ventralized effect. Plants doubly mutant forever and another mutation called *dichotoma (dich)* show a much more complete transformation of all petals to a ventral-like state, resulting in a radially symmetric flower. Not only are these genes required to establish a dorsal-ventral axis of symmetry across the flower, but it is also clear that these genes participate in the asymmetric development of the individual petals along the dorsoventral axis. Although the symmetry-inducing mutations have the most dramatic effect on petal form, they also affect more subtle aspects of zygomorphy in the other organs of the flower. The *CYC* gene is expressed early in floral development in the dorsal part of the flower primordium and then later in the two dorsal petal primordia and the dorsally placed stamenode. *DICH,* like CYC, is expressed in the dorsal part of flower, suggesting that together, these gene products are responsible for establishing dorsal cell fates. These genes also appear to negatively regulate the

DIVARICATA gene, which is responsible for establishing ventral identity in the *Antirrhinum* flower. The patterns of *DICH* and CYC expression, however, do not explain the phenotypic effects of the corresponding mutations on lateral petal development.

It is possible that *DICH* and CYC actually are expressed in the developing lateral petals but below the level of detection, or these genes may act in a nonautonomous fashion to regulate lateral petal identity and form. One possibility is that these genes are involved in stabilising a morphogenetic dorsal-ventral gradient that specifies positional information across the flower and that this information is interpreted independently by each cell in each petal primordium. Surprisingly, the maize gene teosinte-branched (tb) shows remarkable sequence homology to CYC and DICH.

Perhaps the most overt example of a regionalized expression pattern in the petal is the often highly elaborate pattern of pigmentation that occurs in many flowers. These patterns of pigmentation do not follow clonal boundaries, and so some type of intercellular signalling mechanism must also be operating at this level to specify these pigmented domains. One of the major classes of pigments is the anthocyanins, which are found predominantly in the epidermis, and concomitant with this distribution, many of the genes required for anthocyanin biosynthesis are specifically transcriptionally upregulated in petal epidermal cells.

The transcriptional activation of many of these genes in petal epidermal cells occurs during the last cell expansion phase of petal development. Many of the anthocyanin biosynthetic genes appeal-to be coordinately regulated at the transcriptional level. A number of investigations into the control of anthocyanin biosynthesis have been instrumental in beginning to define those gene products that may be essential for the control of re-ion-specific pigmentation patterns. A similar type of regulation has been implicated in the regional activation of anthocyanin biosynthetic genes in *Petunia*. The *Petunia an1* and *an2* genes are required for the appropriate transcriptional activation of a subset of anthocyanin biosynthetic genes, and they encode a basic helix-loop-helix protein and an myb-related gene product, respectively. Although little is known about how these putative transcription factors let to specify the regional activation of anthocyanin biosynthetic genes in this system, at least one component in this signalling pathway has been identified.

The *Petunia an11* gene encodes a cytoplasmically localized WD repeat containing protein that appears to act upstream of the *an2* gene. The an11 gene product has been postulated to regulate the activity of *an2* and other transcription factors involved in anthocyanin biosynthesis by modulating post-translational modifications; required for transcriptional function. It is also possible that the coordinated upregulation of anthocyanin biosynthetic genes is brought about in part by the action of the floral organ identity genes or other MADS box containing genes. CArG box sequences have been found in the promoter regions of a number of anthocyanin biosynthetic genes in *Antirrhinum,* including the *NIVEA* gene, which encodes chalcone synthase. In addition, several petal-specific genes ilia putative zinc finger containing transcription factors have been identified, which also contain CArG boxes in their promoters. These observations suggest at least part of the genetic hierarchy required for regional activation of anthocyanin biosynthetic genes may include a transcriptional regulatory cascade and may be regulated directly by the floral homeotic genes in concert with regional specific cofactors.

These observations suggest that differential transcription factor activity may play a role in the spatial regulation of anthocyanin biosynthesis. Nonetheless, the mechanistic basis by which the anthocyanin biosynthetic genes are activated in spatially restricted domains within the petal is still unclear.

Based on the very dramatic region-specific patterns of pigmentation that arise in cosuppressed plants, a model has been proposed that posits a morphogenetic gradient of transcription factor activity that could serve to specify positional information across the developing petal. Cosuppression occurs when all introduced transgene results in the silencing of both the endogenous gene as well as the introduced copies. The loss of' chalcone synthase expression and concommitant development of unpigmented white petal cells in *Petunia* flowers due to the introduction of a *chalcone synthase A (chsA)* gene, is a well-characterized example of cosuppression. Cosuppression is not unique to chalcone synthase expression but has been observed for a number of other loci, including other anthocyanin biosynthetic genes. However, it does appear that some types of genes may be more prone to silencing, as the chalcone synthase gene *NIVEA* from *Antirrhinum* also is susceptible to cosuppression-like silencing effects. Flowers that display a chalcone synthase cosuppressed phenotype have patches of normal, intensely pigmented epidermal cells adjacent to completely unpigmented cells.

Petal epidermal cells are very regular and conical, and this shape is thought to maximize petal colour by enhancing light absorption by pigments. In *Antirrhinum,* a mutation in the MIXTA gene causes the epidermal cells to develop as flattened rather than cone-shaped cells, resulting in paler colored petals. The MIXTA gene encodes a putative myb domain related transcription factor, and its function is required to generate the specialized, thickened cell walls that give the petal epidermal cells their conical shape. In turn, this conical shape enhances the optical properties of the epidermal cells, in part by focusing light specifically in the pigmented parts of the epidermal cells.

It has been proposed that MIXTA represents a relatively late-acting target gene activated by *DEF and GLO* however, there is, as of yet, no direct evidence supporting this suggestion. Stamens are more complex structures morphologically than petals, with a number of distinct tissue types, many of which are associated with the formation of, pollen. Stamens are composed of an elongated filament topped by an anther, which contains the tissues involved in sporogenesis. The anther is composed of it layer of epidermal tissue surrounding four microsporangia.

The microsporangia are composed of' several cell layers: an outer endothecial cell layer, a middle layer, and the inner tapetum, surrounding the sporogenous tissue, which ultimately will give rise to the pollen. During stamen development, the anther primordia generally become apparent first, followed by the developing filaments. Litter in development, the growth of the filaments is quite rapid and occurs predominantly through cell elongation. In the developing anther primordium, the precursors of the microsporangial tissue increase their rates of cell division, and the cells themselves can also enlarge. The microsporangial cell layers appear to arise from different lineages of cells, depending on the species. Within the sporogenous tissue, internal cells enlarge to become the microsporocytes, which undergo meiosis to produce tetrads of microspores that develop into the pollen grains. Despite the importance of regulating male fertility in hybrid seed production and controlling crop plant fertility, comparatively little work has been done to characterize the genes required for stamen development. Several genes involved in stamen development have been identified either mutationally or by searching for transcripts that are abundant in stamen tissues. Many genes that are highly expressed during stamen development have been isolated on the basis of differential or subtractive hybridization methods. The predominant class of genes isolated in such screens is expressed in the tapetum, and many encode putative secreted proteins. This is consistent with the role of the tapetum in providing nutrients and cell wall components to the developing microspores.

Mutational studies have been carried out in an attempt to identify those genes that are required for different aspects of stamen development. It has been estimated that there are approximately 25,000 genes expressed in a tobacco anther. Many such genes also are expressed in other tissues, making the identification of the corresponding Mutations and the definition of their role in stamen development more problematic. Most Mutational analyses have focused on identifying and characterizing male sterile mutations. Male sterile mutations have been loosely grouped into three classes: structural, functional, and sporogenous.

Structural mutants lack anthers or are otherwise so impaired morphologically that they rarely produce pollen. Functional mutants can produce pollen, but have abnormal morphologies such that the pollen is not released effectively. Sporogenous mutants are morphologically normal, but do not produce functional pollen. Most mutants that have been well characterized are sporogenous male sterile mutants and, thus, are likely to be defective in terminal differentiation processes. Comparatively little work, has been carried out on functional or structural mutants, although such lesions have been identified in several systems. Several structural mutants resulting in male sterility appear to be due to defects in hormone responses. In particular, gibberellin (GA) appears to play a critical role in stamen differentiation in a number of species. For instance, the *stamenless-2* mutation of tomato, which results in short, thin stamens, can be rescued by exogenously added GA. A number of dwarf mutations have been described in maize, which cause the proliferation of stamens' in the normally pistillate ear. Several of these dwarf mutants have been shown to be defective in GA biosynthesis. The *tasselseed2* gene, which is required for the production of the staminate tassel in maize, has been shown to encode a putative steroid dehydrogenase and may be involved in the regulation of the GA-mediated balance between male and female flower development. Other plant hormones, in particular auxins, cytokinins, and brassinosteroids, have also been shown to have an effect on the development of the reproductive organs in some species. Therefore, it appears that the regulation of the relative levels of hormones in particular tissues may have a critical role in stamen differentiation. Male sterility also can be caused by it mitochondrially inherited trait, known as cytoplasmic male sterility, which has been documented in many plant species.

Cytoplasmic male sterility results in defects in anther function and concomitant defects in pollen development. The mitochondrial genes that are affected in cytoplasmic male sterility have been identified in several species. In maize, one form of cytoplasmic male sterility appears to result from a rearranged mitochondrial sequence, which encodes a chimeric polypeptide that spans the inner mitochondrial membrane. In *Petunia,* another chimeric mitochondrial gene produces a polypeptide of unknown function, which is associated with cytoplasmic male sterility in that species.

Mitochondrially induced male sterility in tobacco also is associated with defects in petal development, suggesting that a common metabolic component required for petal and stamen development is perturbed in these lines. Although the mechanism by which altered mitochondrial function can cause cytoplasmic male sterility is unclear, at least one possibility is that any defect in mitochondrial function in the tapetal cells would adversely affect the ability of this tissue to deal with the heightened demands of pollen development. This is consistent with the high numbers of mitochondria that are normally observed in the tapetum.

Despite the wealth of genes involved in stamen development that have been identified by these approaches, little work has been done to establish the mechanisms by which these genes function to produce the complex tissues of the stamen. A few studies have been undertaken to examine whether there are inductive interactions within the tissues of the anther during late stages of development. Stamen-specific gene promoters have been used to drive cytotoxic gene expression in the tapetum to examine the effects of the consequent cell ablation on the development of the pollen.

The ablation of tapetal cells at different stages in development has markedly different effects on fertility. Tapetal cell ablation early at the tetrad stage of pollen development, results in a lack of functional pollen and concomitant sterility. On the other hand, tapetal cell ablation at the later stage of uninucleate microspore development has little effect on fertility. During terminal differentiation of the anther, the pollen is released by a developmentally programmed splitting of the anther along a line of specialized stomium cells. Ablation of the stomium cells has shown that functional stomium cells are required for dehiscence and release of pollen. These observations indicate that temporally and spatially coordinated interactions between the anther cells and the developing pollen grains are critical for insuring male fertility.

One of the clearest examples of how cells can be respecified as to their fate i the phenomenon of floral reversion. Many species can undergo either natural or experimentally induced floral reversion in which flower primordia that are already initiated differentiate as vegetative structures in response to changes day length. Impatiens, which is a short day plan can be induced to revert to vegetative growth with long day treatments. The newly formed leaves that arise are arranged in a typical floral phyllotatic pattern, indicating that morphogenesis but not the inception of organ primordia h been altered. Organs that are initiated at the time reversion can display a partial reversion, with the tip determined before, and the base determined after, the switch to noninductive conditions. In second whorl organs, for stance, that have undergone partial reversion, cells at the boundary can display chimeric features in that individual cells can have characteristics of both petal and leaf. Similarly, partial restoration of the *AP3* homologous *PMADS1* function in *Petunia* can result in chimeric petal-leaf cells.

Therefore, it appears that cells are not committed to a particular fate even up the their last division, and their differentiated fate depends on a combination of cues perceived at a particular time point in development.

The specification of terminally differentiated cell fates also depends on the continued expression of the floral homeotic genes. Several of the floral homeotic genes required for petal and stamen development appear to be required late in development to specify a particular CCII fate. Temperature shift experiments indicate that the *Arabidopsis AP3* gene product is required until the petals and stamens begin to differentiate. Activation of floral homeotic gene expression late in development can also trigger a change in differentiated cell fate. For instance, an unstable *DEF* mutation induced by the Tam transposable element in *Antirrhinum* can revert by excision of the element, restoring *DEF* function. Such excision events can occur so late in development as to result in patches of petal epidermis as small as four cells. Nonetheless, the mosaic patch of petal tissue has the appropriate dorsal and epidermal characteristics or petal cells that would normally arise in that location. This indicates that even at one or two divisions prior to terminal differentiation, cells retain the ability to alter their fate in response to changes in gene expression.

Chapter 9

DNA Technology and Ecological Evolution

DNA as the genetic material and the determination of its structure were all made in the absence of any way of directly confirming the central tenet of the new genetics. There was ample indirect evidence to prove that the informational content of DNA lies in the order of its nucleotide subunits, and that this is reflected in the amino acid sequence of the proteins it specifies.

The biological revolution that led to the new genetic technology began very quietly with the study of an unusual bacterial defense system known as restriction and modification. Foreign DNA can arrive in a cell by a variety of means such as genetic transformation, transduction, transfection, conjugation, and even simple viral infection. However, if this DNA is not homologous to the host cell (that is, from the same or closely related species), it is not necessarily to the long-term advantage of the host species to allow it either to recombine or to persist as a self-replicating plasmid within the cytoplasm. Homology during recombination is a host cell's first line of defense; however, even if the foreign DNA fails to recombine, it may linger within the cell. The means by which a host cell eliminates most foreign DNA actually involves two DNA site-specific enzyme activities. In each case, the enzyme recognizes the same base sequence on the DNA and binds to it. In the normal course of events within a cell, the first or *modification enzyme* adds a methyl group to a certain base within the recognition sequence during DNA replication. This modification does not seem to affect the coding properties of the DNA. The second enzyme, a *restriction endonuclease,* will not bind to a modified DNA sequence, however, although it will bind readily to an unmodified one. The type of enzymes a generally used in DNA technology are the separate variety. The various restriction endonucleases are designated according to the first letters of the genus and species name of the producing organism followed by a roman numeral used to identify separate enzymes within the same organism. For example, *Anabaena subcylindrica* produces the enzymes *Asu*I, *Asu*II, and *Asu*III. If the information coding for the restriction and modification enzyme is located on a plasmid, that too can be noted. *Eco*RI is an enzyme produced by *E. coli* cells that carry the RY 13 plasmid. In other cases, the final letter can indicate the strain (a variant within species) that is the producing organism. Although the binding of restriction endonucleases specific, the cuts made by them are not necessarily so. A type I enzyme such as *Eco*B cuts nonspecifically at some distance from the binding site (in this case about 1,000 base pairs). By contrast, a type if enzyme such as *Eco*RI both binds specifically and cuts, specifically. It is this latter type of enzyme that is so interesting from the standpoint of DNA technology.

A few type III enzymes such as *Eco*PI are known to have properties intermediate between types I and II. They bind to a specific DNA sequence but make their cuts a specific distance away from the binding site; therefore, the position of the cut can be predicted, but the sequence to be cut cannot.

In several different studies by Nobel Prize winners Werner Arber, Daniel Nathans, Hamilton O. Smith, and their various collaborators, it was shown that DNA fragments produced were of two types. They might have blunt ends (for example, *Hpa* I) or overlapping single-strand ends (*Eco*RI). Of critical importance is the fact that all DNA fragments produced by a particular type II restriction endonuclease have absolutely identical ends regardless of the source of the DNA.

DNA fragments can be brought together and the single-strand ends allowed to pair or reanneal. Random combinations of linear DNA molecules held together by hydrogen bonds result. As the molecules increase in length, however, they have a tendency to fold back on themselves so that they circularize. If these circular molecules have their nicks sealed by the same DNA ligase enzyme used for replication and if they have an appropriate origin of replication, new plasmids will have been formed. There are several important points to note about DNA splicing reactions.

Clearly, fragments produced by any restriction endonuclease can be used. If the fragments have blunt ends, however, highly concentrated DNA solutions are a necessity, and DNA ligase must be present during the mixing step in order to join fragments together as they collide. When fragments with cuts at specific base sequences do join, the recognition site for the restriction endonuclease is regenerated; that is, DNA fragments that have been spliced together can be cut apart later by the original enzyme to reform the original fragments. Joining of DNA fragments results both in random combinations of the fragments and in random orientation of the fragments within the growing molecules. However, DNA splicing occurs only in a test tube (in vitro) and requires no real base sequence homology except in the sense that all restriction fragments produced by a particular type II enzyme have identical ends. On the other hand, genetic recombination occurs in a living cell (in vivo) and usually requires substantial sequence homology throughout the region. For the purposes of this text, therefore, the terms *recombination* and *recombinant* are reserved for the natural process, and the in vitro process is referred to as DNA splicing or gene splicing. Genetic engineering can refer to organismal crosses either from true recombination or from DNA splicing. In the modern vernacular, however, the term *genetic engineering* usually refers to DNA splicing. In order for spliced DNA to be maintained stably within a cell, it must be able to replicate. Replicational functions might be provided if the spliced DNA is made to recombine with the host cell genome so as to integrate the new DNA into the host replicon. Vectors can be any temperate virus, resistance plasmid, bacteriocin plasmid, etc. that is capable of replication within the desired host cell. The ideal vector also carries some sort of selectable marker (gene) so that its presence in a cell can be detected easily. One type of selectable marker commonly used is an antibiotic-resistance gene from an R plasmid.

Plasmids carrying spliced DNA normally can be introduced into cells by genetic transformation. In some cases, it is possible to carry out in vitro packaging of viral DNA into protein coats, and in such instances, fully infectious transducing viral particles are formed. Whatever the means used to introduce spliced DNA into a cell, once that cell has begun to replicate the DNA and undergo cell division, the entire clone of cells that it generates carries the spliced DNA. This spliced DNA is said to have been *cloned*.

DNA replication is sufficiently different that plasmids able to replicate well in a bacterium are unable to do so in a eukaryotic cell unless something like a viral origin of replication is added. Yeasts, however, have several plasmids that normally occur in their cells and can serve as good vectors. For more complex organisms, the vectors of choice usually are either transposons or viruses. The most frequently used viruses are retroviruses, because of their integration into the host chromosome, and vaccinia virus, because it has many nonessential genes that can be deleted to make room for spliced DNA. The most common method of getting spliced DNA into a eukaryotic cell is genetic transformation, although transduction is also possible. Sometimes it might be desirable to perform most of the genetic manipulations on a piece of DNA in a prokaryotic cell and then transfer it into a eukaryotic cell to study its expression. This feat can be accomplished with a *shuttle vector* that carries two origins of replication, one for the prokaryote and one for the eukaryote.

Successful *cloning* of a spliced DNA molecule does not, however, guarantee that the inserted DNA fragment will be expressed in a manner that will alter the phenotype of its host cell. For this reason, vectors usually carry one or more selectable genetic markers, as mentioned previously. In the case of a vector with two or more markers, it is frequently advantageous to splice the extra DNA into the middle of the coding sequence of one selectable marker so as to create an insertion mutation that can inactivate the coding sequence. Then, the presence of one genetic marker and the absence of another can be taken as evidence that the cell is carrying spliced DNA. However, DNA fragments inserted into the replicational region act to prevent the plasmid from replicating; therefore, it would be lost. Insertion of a DNA fragment coding for a protein into a site, such as *Pst*I, that is downstream from a promoter results in the formation of a *fusion polypeptide*, a protein having the amino terminus of the β-lactamase enzyme (an inactivator of ampicillin) attached to the amino end of the protein encoded on the cloned DNA.

Fusion polypeptides may or may not function as enzymes, depending on the exact nature of the fusion. In order for a spliced DNA fragment to express its function uniquely, therefore, it is generally necessary for the fusion polypeptide to be separated into discrete proteins or for transcription to occur in a manner that separates cloned DNA from vector DNA. Transcription of this nature means that a promoter site must be provided either on the spliced fragment itself or on the neighbouring region of the vector. In addition, if a protein is to be produced, it is necessary for the appropriate ribosomal binding site to be present after the promoter.

As the properties of restriction enzymes became clear, there was widespread debate within the scientific community over the extent to which the use of these enzymes presented a biological hazard. The general theme of the arguments raised revolved around how DNA splicing could be used to circumvent the so-called "species barrier" to genetic exchange. It was thought that once this barrier was broken, there might be wholesale rearrangements of genetic traits and inadvertent production of genetic "monsters" that could have serious effects on other members of the biosphere. Clearly, such a concern was justified based on the information available at that time. The conversion of a disease-causing bacterium into a form resistant to the commonly used therapeutic antibiotics would be of grave concern. Similarly, any improvement in the ability of such a bacterium to cause disease would be undesirable indeed. Even more elaborate scenarios were envisioned in which combining segments of eukaryotic and prokaryotic DNA would lead to entirely unexpected results such as an *E. coli* cell that carried DNA capable of stimulating tumour production in animals.

Scientists involved in this research were so concerned that they met in Asilomar, California, declared a moratorium on DNA splicing research until such time as appropriate safety guidelines could be developed. Stringent guidelines were prepared and implemented under the auspices of the National Institutes of Health and its new component, the Recombinant DNA Advisory Committee. In recent years, this committee has relaxed the guidelines significantly as it has become evident that early fears were vastly overestimated. In part, this overestimation was due to a lack of specific knowledge regarding the considerable difficulties with which eukaryotic DNA can be expressed in prokaryotes, as indicated previously. We also now know that transfer of DNA from prokaryotic to eukaryotic cells is promoted routinely by the Ti plasmid.

Additionally, careful examination of the older literature as well as contemporary experimentation, has indicated that prokaryotes are capable of biochemical feats once thought to be the sole province of eukaryotic cells. For example, bacteria that spontaneously produce substances that mimic the activity of insulin, progesterone, and other animal hormones have been observed. The guidelines for safe use of spliced DNA are adapted from the principles outlined for the containment of disease-causing microorganisms. These guidelines address both the biological and physical aspects of spliced DNA containment. Physical containment guidelines are taken directly from the Centers for Disease Control biosafety rules. They are easier to under stand because the basic idea is to reduce the probability of an organism that carries spliced DNA escaping from the laboratory. All laboratory facilities are rated on a four point scale, a BL1 facility (biosafety level 1) being the least contained and a BL4 facility being the most contained. Previously, a BL1 facility was called P1, and BL4 was designated P4, etc.

A BL1 laboratory is a typical well-maintained microbiology laboratory in which lab coats are worn;

eating, drinking, and smoking are prohibited; and hand washing facilities are available. In addition to these same features, a BL2 facility has limited access to approved personnel only. A BL3 facility is not only limited in access but also has a double door arrangement to aid in preventing any contamination from escaping. Much of the work in a BL3 facility must be performed in biological safety cabinets that filter the air before recirculating it to a room.

The ultimate or BL4 facility is one in which self-contained anticontamination suits or glove boxes are used for all manipulations. Reduced air pressure relative to the external environment is maintained at all times; therefore, when a door is opened, air rushes into the building. The problem of biological containment is analogous to physical containment, but the object is to prevent cloned DNA from leaving its host cell and escaping into an unwanted or undesirable host. If the vector used in the cloning is a conjugative plasmid like F or a normal virus like phage λ, the probability of the vector's escaping the host cell is high, provided that the host can survive outside the laboratory environment for a reasonable period of time.

These considerations are taken into account in the creation of host-vector systems for cloning. If the host is highly viable, the vector is weakened and vice versa. With experiments perceived as potentially dangerous, certain levels of biological containment are required. For example, in cloning experiments involving *E. coli* as the host cell, an EK1 host-vector system employs a normal laboratory strain of the bacterium, while an EK2 host-vector system utilizes a specially weakened strain that cannot grow on naturally occurring amino acids. The weakened strain survives only in the research laboratory, therefore. Biological and physical containment are complementary systems; therefore, an increase in the level of one allows for a decrease in the level of the other. Some specific cases are listed in Table 1 to illustrate how biological and physical containment are used together in a laboratory setting.

Blotting techniques

Many blotting techniques have been developed in which macromolecules are separated by *gel electrophoresis* first. Most often a gellike support composed of acrylamide (small pore size) or agarose (large pore size) is used. A sample of protein or nucleic acid is placed at one end of the gel to which an electric field is applied, forcing the charged molecules to migrate toward the appropriately charged electrode. The rate of migration is affected by the size and shape of the molecules.

The position of the macromolecules on the gel can be detected by the use of dyes such as Coomassie blue for protein or ethidium bromide for nucleic acids. A general dye, however, shows the position of all macromolecules, not just the one in which the experimenter is interested. This problem requires for its solution a method in which a protein or nucleic acid that has a specific sequence of amino acids or bases can be recognized. After the molecules have been separated, they can be blotted or transferred sideways to a paper-thin matrix that has been laid on top of the gel. This procedure gives a durable substrate that can be handled with less fear of breaking. Originally, the matrix material used for transferring nucleic acids was nitrocellulose, but more recently-certain charged nylon derivatives have been employed.

Once the macro-molecules are attached firmly to the matrix, various techniques can be used to detect the position of the molecules of interest. Essentially similar procedures can be used to trap nucleic acids or proteins within individual bacterial colonies or phage plaques.

The blotting material is laid carefully on an agar plate so that cells or phages can adsorb to it. The matrix is treated to break open the cells or viral particles and then washed to remove unbound material. For nucleic acids, the simplest sequence-specific detection method is to use a single-strand DNA *probe* that has been labelled in some fashion, usually with ^{32}P.

After washing, the solid support is tested for the presence of radioactivity. If there is no homology between probe and bound nucleic acid, the radioactive label is lost. If there is perfect homology, the probe binds strongly, and the regions to which it has bound can be detected by autoradiography. If there is imperfect homology, the result depends on the stringency of the incubation conditions. High pH, high temperature, or high concentrations of formamide all tend to break hydrogen bonds. Probing under any of these conditions requires better sequence homology than probing under less stringent conditions. Some laboratory workers prefer to use nonradioactive methods of detecting DNA probes. In one such method, the biotin-avidin system is employed.

The vitamin biotin binds exceptionally strongly to avidin and forms a very stable complex. It is possible to attach biotin derivatives to nucleic acids in such a way that the nucleic acids' ability to form hydrogen bonds is not affected. When in a duplex, the biotin groups on the DNA are still accessible to avidin. The avidin protein itself is joined covalently to an enzyme, such as alkaline phosphatase. After the blotting process is completed, the substrate for alkaline phosphatase is washed across the surface of the solid support. Wherever the avidin has bound, a coloured compound is formed. As with autoradiograms, the presence of a dark spot on the stained blot indicates binding of the probe. The initial blotting technique was developed with single-strand

DNA probes in the laboratory of Earl Southern and quickly became known as *Southern blotting*. Of course, there is no reason why RNA cannot be used also. Because the opposite type of nucleic acid is used, this technique is known as *Northern blotting*.

The polymerase chain reaction (PCR) is the long-awaited solution to the problem of detecting rare DNA sequences within a heterogeneous population of DNA. In order for the technique to work, it is necessary for the experimenter to have two DNA fragments that contain a short sequence of DNA known to flank or be part of the region of interest. In the case of cloned DNA, these-fragments can simply be portions of the vector DNA itself that are located at each end of the cloning site. The two fragments must be from opposite strands. The experimenter sets up an in vitro DNA replicational system using the fragments as primers. If the temperature is raised briefly, the DNA fragments can be denatured. DNA polymerase is added at this time so that a leading strand of DNA can be synthesized using an added DNA fragment as a primer. The corresponding lagging strand is not replicated in this system. Because the primers flank the region of interest, however, each DNA strand is copied. In the region in which there originally was one double-strand DNA molecule carrying the sequence of interest, there are now two. Reheating of the mixture denatures both molecules.

New primers and enzymes then can be added, and the number of available molecules ought to double again. This chain reaction derives its name from the repeated cycles of DNA replication. One problem with the PCR technique is that each time DNA fragments are denatured by heating, the activity of the DNA polymerase enzyme also is destroyed. A major breakthrough in using this technique came when DNA polymerase from the extreme thermophile *Thermus aquaticus* was isolated. This enzyme, called *Taq* polymerase, is stable at temperatures that denature DNA; therefore, the enzyme need be added only at the beginning of the chain reaction. Automatic equipment is now available to take a small sample through a series of 20 to 30 cycles, each time amplifying the desired sequence approximately by a factor of two.

The genetic technology described in the preceding sections has effects on all facets of microbiology. In subsequent chapters, the information presented in this chapter is enlarged upon and used to illustrate specific applications for the products of genetic engineering. The technology involved has already become a part of routine microbiology in taxonomic studies and in diagnostic laboratories. From the point of view of the diagnostician, genetic engineering provides a major technological improvement, the ability to detect rare macromolecules within a heterogeneous population. For example, the standard test to confirm whether an individual is infected with the HIV-1 virus that causes AIDS is a Western blot that detects specific viral proteins within the sample of patient blood.

Specimen identification of pathogenic bacteria often is performed with Southern blots and DNA probes specific to the particular genus or species. The most recent advance is in the realm of PCR technology. In theory, an appropriate PCR should be able to amplify a single molecule of DNA to the point at which it can be detected visually on an electrophoretic gel. The only real problem is to locate DNA sequences that can be used as flanking primers for the synthetic reaction. Sequences chosen must be highly specific to the organism(s) of interest so that false positive reactions do not occur.

The reaction itself requires part of one day to run, and DNA analysis adds only a few more days to the assay. For a slowly growing organism, therefore, identification may be possible in the amount of time that formerly was spent in waiting for a colony to appear on an agar plate. A theoretical limitation to the assay of the pathogens in question physically present in the laboratory sample.

As with any new technology, genetic engineering has presented a number of difficult issues that need to be resolved over the next few years as we attempt to apply present-day knowledge and knowledge to be acquired in the foreseeable future. In this section, just a few of these issues are surveyed, but no solutions are outlined because there are no easy solutions. Instead, you are urged to keep these problems in mind and consider what are appropriate and inappropriate uses of DNA splicing technology.

One significant regulatory issue that has arisen concerns the use of bacteria in the environment. Should bacteria be genetically engineered for environmental use? There are many separate questions that require answers. Among them are:

1. Is it appropriate for a company to produce large quantities of a normal bacterium for release into an area in which it is not normally found? If the answer is no, some potentially important activities like the use of bacteria to degrade environmental compounds are eliminated.

2. If we assume that releasing naturally occurring bacteria in new habitats is acceptable, is it appropriate to do the same thing with genetically engineered bacteria? It the answer to that question is no, how are the situations different?

3. If genetically engineered bacteria are released into the environment, what sorts of risks is society taking

in the process? What kinds of information should be provided about the bacterium before it can be released?

Questions like these are presently before the U.S. Environmental Protection Agency. Over the course of the next few years, answers to these and a number of other questions will have to be given. Presently, the general tendency is to treat bacteria like new chemicals to rule that bacteria altered in any way other than by simple deletion mutations in the laboratory must be tested for safety before they are released.

The regulatory process already has engendered complex lawsuits as companies with marketable products clash with those concerned about the potential effects of large numbers of novel bacteria displacing those present normally in the environment. Turning to a different dilemma, consider the ethical questions that can be raised by the therapeutic uses of genetic engineering. Human interferon can be produced in large quantity. Is the use of this protein in therapy much like any other drug, or is it not? What safety and efficacy tests should be applied to biochemicals produced by DNA splicing? If the cloned interferon is the same as that found in the human body, no unknown molecules have been introduced if the recombinant interferon is administered.

Currently, if the recombinant protein is 100% identical to the naturally occurring one, a manufacturer only has to prove its purity and safety. What if a modified protein can be shown to work better? How much testing should be required before that protein is allowed on the market? This is a very important question because manufacturers spend tens of millions of dollars to run the full battery of tests needed to bring a new drug onto the market. Finally, consider what happens as tools become available to treat the causes of various genetic diseases. Suppose that an individual is ill due to the lack of a particular enzyme in her body. If vaccinia virus was used as a vector and DNA that coded for the missing enzyme was spliced into the virus, it might be possible to infect the sick individual with the recombinant virus. If the viral infection was successful, and if the spliced DNA integrated in the chromosome, and if the inserted DNA was expressed properly, the sick individual might be cured. Is the treatment still appropriate? Suppose you had a way to introduce that virus into the germline cells so that all the offspring of that individual would inherit the virus and the spliced DNA. Is that sort of experiment appropriate? Should some sort of line be drawn for experiments on humans? If so, where? Who should draw such a line? What effects might we have on ourselves as a species if we start to do these kinds of experiments? These are important questions to which society must find answers over the next few years.

The nature of the genetic code, the existence and role of messenger RNA, the mechanism of translation and the basic principles of gene regulation in bacteria were all established indirectly, without being able to determine the nucleotide sequence of the DNA involved. The problem lies in the great size of DNA molecules, and the chemical similarity of one DNA molecule to another. The DNA molecules from the chromosomes of eukaryotes are millions of base pairs long and each contains thousands of different genes.

Although each DNA is quite unique in its informational content, from a chemist's point of view one DNA molecule is much like another, a featureless string of only four chemically distinguishable subunits repeated at random many thousands of times. It presents no unique landmarks that can be used to pull out particular pieces of DNA. Another aspect of the problem is that the

DNA representing a typical protein-coding gene is present in only a few copies per cell and represents an infinitesimally small fraction of a sample of DNA from plant or animal tissue, or even from a bacterial culture. Even if it had been possible to isolate a particular piece of DNA, to obtain any plant or animal gene in sufficient amounts for analysis seemed an impossibility.

Ways were needed to isolate identifiable DNA fragments of manageable length in sufficient quantity for chemical analysis. By the beginning of the 1970s the tools were at last to hand in the form of an exceptional group of DNA-cutting enzymes from bacteria—the restriction enzymes—which snip out particular pieces of DNA from the heterogeneous mixtures of DNA molecules in natural DNA samples. Restriction enzymes were also used to construct recombinant DNAs from these fragments, in which form they could be multiplied to give sufficient DNA for analysis.

These techniques have transformed basic research, as well as forming the basis for a multimillion pound biotechnology industry. Artificial recombinant DNA is a DNA that has been constructed outside a living cell by joining two or more pieces of DNA from different sources to produce a novel combination of genes. It usually, but not always, involves a combination of DNA from widely different species, such as the splicing of human or animal genes into a bacterial or yeast DNA.

These techniques enable small fragments of DNA to be introduced into bacterial or yeast cells as part of a bacterial or yeast DNA, where it will then be replicated under the control of the host cell to provide a virtually unlimited source of cloned copies of that DNA. This is an essential step in isolating individual genes from the large genomes of plants and animals. With further manipulation, microorganisms can be persuaded to express the foreign genes they carry, to mass-produce medically and commercially useful proteins that are difficult or impossible to obtain by any other means. Pure cloned DNA can be introduced into plant and animal cells, including the fertilized eggs and early embryos of mammals, and, from these, transgenic organisms can be raised that carry potentially useful genes from a wide variety of species. Behind these well-publicized feats of genetic engineering lies a wide range of techniques for isolating, identifying and altering individual pieces of DNA that have been developed during the past 15 years and now go under the collective title of recombinant DNA techniques or recombinant DNA technology.

Cloned DNA can be altered to order *in vitro* by novel mutational techniques and then reintroduced into cells to be translated into new proteins with slightly altered properties. This is the basis of the emerging art of protein engineering, the second generation development of recombinant DNA technology. This technique is predicted eventually to produce novel enzymes and other proteins of industrial or medical importance to order, and to create 'hybrid' proteins with quite new properties—such as 'enzymatic' antibodies that combine in one molecule an enzyme activity and an antibody's exquisite specificity for a particular target.

Mutation to order has also proved a powerful technique for studying the ways in which gene expression is regulated in the cells of multicellular organisms. Recombinant DNA technology has made possible some completely new approaches to old problems—such as the mapping of gene locations on the chromosomes and the study and detection of inherited genetic disease. As well as the topics in this section, some uses of recombinant DNA techniques in cancer research, immunology and in developmental biology are also mentioned elsewhere.

One of the most useful aspects of recombinant DNA to biologists in all fields is the ability it gives to produce large amounts of rare or hard-to-purify proteins by translation of cloned genes, and it is also now easier in most cases to determine the amino acid sequence of a protein by working backwards from the nucleotide sequence of the DNA that specifies it. Microorganisms into which novel DNA has been introduced are often referred to as recombinant microorganisms and the proteins produced from such recombinant organisms as recombinant insulin, recombinant interferon, etc. to distinguish them from the same proteins isolated directly from animal tissue.

Recombinant DNA is often abbreviated as rDNA. (As this is also a long-standing abbreviation tor the DNA specifying ribosomal RNAs, the abbreviation rtDNA has also been suggested to avoid confusion.) In a wider sense, the terms recombinant DNA and recombinant organism are now also applied to any DNA that has been altered by direct manipulation outside the cell, and to any organism into which that DNA is introduced, even if the same effects might have been achieved (given infinite patience and good luck) by the natural processes of mutation and genetic recombination.

This distinction is not simply of academic importance. In most countries, there are regulations governing recombinant DNA work over and above those that relate to the usual handling of microorganisms or release of 'natural' organisms to the environment for biological control purpose's. Whether an organism is classed as a 'recombinant' genetically engineered organism, or one bred by 'natural' means can make a great difference to both bureaucracy's and the public's perception of the comparative 'hazards' of such organisms even when the final result is biologically virtually identical.

Restriction enzymes, more correctly called restriction endonucleases, are the molecular scalpels without which none of today's powerful techniques for analysing and manipulating DNA would be possible. They are a group of exceptional DNA-cutting enzymes from bacteria, which differ from other nucleases in only cutting a DNA chain at a precise point within short, precisely defined nucleotide sequences (restriction sites).

DNA molecules are extremely long and chemically very featureless, being made up of only four different subunits repeated many thousands of times in a chemically random sequence. This is a formidable obstacle to isolating a particular fragment of DNA or determining its sequence. Any attempt to cut a DNA preparation from natural sources into more manageable pieces by most chemical and enzymatic treatments leads to a mixture of anonymous random fragments, which it is impossible to analyse further. The restriction sites that occur by chance in all DNAs provide natural chemical guideposts along the DNA. Because most restriction sites are only a few base pairs long they will occur in any DNA of sufficient length, whatever its origin.

In the DNA of the small animal virus SV40, for example, the restriction site for the endonuclease EcoRI (a restriction enzyme from *Escherichia coli*) occurs once, that for the enzyme HpaI (from *Haemophilus parainfluenzae*) four times and that for the enzyme HindIII (from *Haemophilus influenzae*) eleven times. A particular restriction enzyme will therefore always cut a given DNA at the same point (or points), generating a set of specific DNA fragments (restriction fragments) that can be sized and separated by gel electrophoresis. Electrophoresis is a general technique for separating DNA fragments or proteins on the basis of their movement through a sheet of polyacrylamide or agarose gel in an electric field. The distance they move in a given time depends on their size and electric charge. Smaller DNA fragments move quicker than larger ones and so a mixture of fragments of different lengths separates out into discrete bands in the gel, each band consisting of fragments of the same length.

Each band therefore consists of a purified source of identical DNA fragments. These can be accurately sized by comparison with the movement of fragments of known length (standards), and on suitable gels fragments differing in length by only one nucleotide can be separated. Conventional gel electrophoresis is used to separate fragments of DNA tens to thousands of base pairs long. New developments (pulsed-field gel electrophoresis) now also allow much larger DNAs of chromosome length—millions of base pairs long—to be separated. Restriction fragments from any source can be inserted into a suitable carrier or vector DNA to produce an artificial recombinant DNA that is multiplied in rapidly dividing bacterial cells to produce a virtually unlimited supply of a given piece of DNA.

Some restriction enzymes simply cut the two strands of the DNA straight across at the centre of symmetry. Others, however, make a staggered cut in the DNA, cutting each strand symmetrically with respect to the other and generating in the process two DNA fragments which have complementary single-stranded ends. BamHI, whose restriction site is illustrated here, is an enzyme of this type, cutting

↓
5' . . . G G A T C C . . . 3'
3' . . . C C T A G G . . . 5'

Because the nucleotide sequence on either strand is symmetrical this generates two pieces of DNA with 'sticky ends'. If two *different* DNA molecules are cut in this way, the pieces will rejoin to each other, creating a 'hybrid' or 'chimaeric' DNA. The source of the DNA is immaterial, DNA from species as far apart as bacteria and humans can be joined up in this way. This feature of restriction enzyme cutting was used to generate the first recombinant DNAs.

DNA cloning is a means of isolating individual DNA fragments from a mixture, and multiplying each fragment to produce enough material for analysis and further use. DNA fragments are enzymatically spliced *in vitro* into small carrier DNA molecules to produce a recombinant DNA. The carrier DNAs or cloning vectors are derived from small self-replicating DNA molecules such as the plasmids and phages that naturally infest bacterial cells. Once inserted into such a carrier, the 'foreign' DNA fragment can be introduced into a suitable host cell, where it will be maintained and replicated along with the carrier DNA. For routine DNA cloning to initially isolate and amplify a gene, specially bred strains of *Escherichia coli* are used as hosts.

Recombinant constructs containing a foreign gene are introduced into bacterial cells by transformation. Bacterial cells naturally take up DNA added to their culture medium and, if the DNA is compatible, they will express the genes it contains, acquiring new properties. On its own, however, a fragment of mammalian or non-bacterial DNA will not replicate or be maintained in the bacterial cell as it does not contain replication and other regulatory signals recognizable by the bacterium, and is soon broken down by bacterial nucleases. Only when incorporated into a bacterial carrier containing the appropriate signals can the DNA be maintained and replicated.

The mechanism of transformation is still something of a mystery as a large molecule such as DNA should not theoretically be able to penetrate the cell membrane. Those bacteria that have taken up the added DNA are selected by the new properties they express; in recombinant DNA work an antibiotic resistance marker gene is commonly incorporated into the vector, so that only those bacteria that have taken up the recombinant plasmid can grow on medium containing the appropriate antibiotic.

DNA libraries are collections of cloned DNAs. There are two types of library—genomic libraries representing the entire genetic complement (genome) of individual animals, plants, viruses or bacteria, and cDNA libraries representing the complete mRNA complement of a single type of cell. Genomic libraries are usually constructed by randomly shearing chromosomal DNA into suitably sized fragments (of around 20 000-40 000 base pairs) either by restriction enzymes or mechanical means, and then cloning the individual fragments into a suitable cloning vector, usually a specially modified phage lambda as the DNA is more easily stored in this form.

The random cloning of an entire genome is often called shotgunning. To ensure that the whole genome has been incorporated into recombinant DNAs, a typical library of a human genome (3.5×10^6 base pairs) will contain between a quarter and half a million overlapping separate clones. Once a DNA library has been established, clones containing a gene of interest can be identified by screening the collection by hybridization with specific nucleic acid probes.

Modern equipment now makes it possible to screen large numbers of clones relatively rapidly, an important practical consideration as hundreds of thousands of clones may have to be screened to pick out ones containing a single-copy gene from a, human genomic library.

The overlap between individual clones is crucial to techniques such as chromosome walking which allows one to proceed from a known site on the chromosome to an unidentified gene nearby for which no specific probe is yet available. This is the situation for many genes implicated in inherited genetic defects whose gene product and function are totally unknown, but whose approximate position on the genetic map has been established by classical genetic methods.

Recombinant DNA work often makes use of synthetic DNA, short polynucleotide chains built up chemically nucleotide by nucleotide in any sequence required and then converted to a double-stranded DNA by enymatic replication *in vitro.* The chemical procedure has now been automated for DNA chains up to 100 nuclecitides or so. Synthetic DNA is sometimes used in place of naturally derived DNA to direct the synthesis of short peptides in genetically-engineered bacteria. A synthetic gene for the small peptide hormone somatostatin was used in the first genetically engineered *E. coli* to produce a mammalian protein. Short synthetic oligonuclecitides are also used to introduce restriction sites or mutations at predetermined positions in a DNA.

Whether isolated as bands on a gel or as cloned recombinant DNAs, restriction fragments are still anonymous pieces of DNA. The most common way of identifying them is to screen the gel or the collection of clones with specific nucleic acid probes that recognize and pick out only a particular DNA sequence. Probes are short lengths of DNA or RNA of known sequence or origin, labelled with a radioactive or chemical tag for easy detection of the DNAs to which they attach. The probes recognize their target DNAs by the exquisitely specific phenomenon of nucleic acid hybridization: this is where a single strand of DNA or RNA pairs up with an exactly complementary strand to form a correctly base-paired double-stranded molecule. The two strands of the DNA double helix are held together by relatively weak hydrogen bonds between the bases of opposite strands. These weak bonds allow the strands to separate easily so that DNA can replicate and be transcribed. In the cell, strand separation relies on enzymes and other special proteins, but if isolated DNA is gently heated or subjected to certain chemical treatments this is sufficient to disrupt the hydrogen bonding and the two strands fall apart. Single-stranded DNA will reform a double helix only by base-pairing with any DNA or RNA of exactly complementary sequence.

Nucleic acid hybridization is central to recombinant DNA work by providing the means of recognizing and isolating a given piece of DNA from a mixture of fragments. A routine method of identifying restriction fragments is illustrated in Fig. 4.3. DNA hybridization is so precise that clones containing a required gene can be picked out from libraries of hundreds of thousands of DNA clones. *In situ* hybridization techniques are used to identify the position of a particular DNA sequence on a chromosome, or the spatial and temporal pattern of specific mRNA synthesis in tissues.

Radioactively labelled DNA or RNA probes are applied to preparations of specially prepared chromosomes, tissue samples, embryos, etc. and the radioactively labelled regions can then be distinguished. The development of probes for identification is central to all recombinant DNA work, and is the most difficult step in most applications. The vicious circle of having to use one nucleic acid to identify another was initially broken in early recombinant DNA research by using as the first probes unusually abundant mRNAs that predominate in, for example, cells specialized to produce a particular protein. These could be isolated in a sufficiently pure, radioactively labelled form by the methods of the day.

The very first genes isolated from eukaryotic cells were therefore those for the globins (constituents of haemoglobin) and ovalbumin (produced in large amounts by chick oviduct cells) and genes from viruses that take over the cell's machinery to synthesize large amounts of their own mRNAs. With the increasing sophistication of recombinant DNA techniques and the exponential accumulation of data on the organization and location of genes along eukaryotic chromosomes, many technical tricks have been devised to identify. DNAs. The most usual procedure nowadays for isolating protein-coding genes is to purify just sufficient of the protein to determine its first 30 or so amino acids by modern microsequencing methods that require only picogram (10^{-12} g) quantities of protein, and then to construct a set of synthetic DNA probes (by chemical DNA synthesis) corresponding to that sequence.

Because of the redundancy of the genetic code there are several alternative nucleotide sequences that could correspond to the amino acid sequence. One of these will correspond to the genomic DNA sequence and will pick out the relevant clone from a library of cloned DNAs. Even a partial sequence is usually selective enough to pull out only the required clone. For very large genes, where the coding sequence may be spread across several clones, the overlap between clones can be used to pull out the next one by using each clone as a probe for the next.

When a gene of interest is isolated and cloned from one species it can then often be identified in other species using the cloned DNA as the probe. Even though the DNA sequences of corresponding genes from even quite closely related species are unlikely to be identical, the conditions in which the probe is hybridized can be 'relaxed' so that it picks out closely related but not identical sequences. Relaxed hybridization is similarly used to screen a library for related but not identical sequences within the same genome, often identifying hitherto undiscovered members of a gene family.

Restriction sites provide the landmarks for finding one's way around a DNA molecule. They allow a newly isolated DNA to be characterized and located in its wider DNA-context, and they are also invaluable for distinguishing one DNA from another. The first step in characterizing an isolated DNA fragment is to construct a physical map (a restriction map) of the different restriction sites along its length. A restriction map is pieced together by digesting portions of the DNA sample with different restriction enzymes. Each digestion produces a typical set of restriction fragments of different lengths. Those covering the area of interest are identified by an appropriate nucleic acid probe. Fragments obtained by digestion with one enzyme are then digested with one or more of the other restriction enzymes (and vice versa) to obtain the location of different restriction sites relative to each other.

Restriction analysis has also been used to trace the ancestry of present-day human populations. The tiny circular DNA molecules contained inside the mitochondria of human cells provide a novel means of tracing the descent and relationships of human populations. Unlike the chromosomes of the cell nucleus, which are inherited from both parents and are also the product of extensive reshuffling by recombination in preceding generations, mitochondrial DNA is inherited only from the mother, through the mitochondria present in the cytoplasm of the egg, which replicate themselves and their DNA as the cells of the developing embryo grow and divide. Mitochondrial DNA cannot undergo genetic recombination and is therefore inherited unchanged through the female line for many generations. Throughout time, however, mutations inevitably accumulate so that the mitochondrial DNAs of unrelated individuals now differ in nucleotide sequence (particularly in non-coding regions) and these differences can be detected by restriction analysis. Comparisons of restriction patterns of mitochondrial DNAs from large numbers of people from different ethnic groups is therefore yet another way of gaining information on human ancestry, origins and dispersal. Interpretation of the results follows the general principle that DNAs with very similar restriction patterns have diverged more recently than those which are less similar, and that population groups with the greatest amount of variation within them are probably of more ancient origin.

A pioneering study of human mitochondrial DNA carried out in recent years focused on people representing five broad geographic regions: sub-Saharan Africans (represented in the study chiefly by black Americans), Asians (China and the Far East), Caucasians (Europe, North Africa and the Middle East), aboriginal Australians and aboriginal New Guineans. The first results provide another strand of evidence to add to palaeontological evidence for an exclusively African origin for present-day humans. When the mitochondrial DNAs are grouped by degree of similarity they form two main lines of descent, both stemming from Africa, one containing exclusively Africans and the other some Africans and all other groups.

More speculatively, the researchers reconstructed a hypothetical 'ancestral' human mitochondrial DNA, and suggested the existence of a mitochondrial 'Eve', who, they calculate, could have lived in Africa some 200 000 years ago and from whom we are all descended. The final step in analysing any DNA is to determine the linear order of its component nucleotides—variously termed the DNA sequence, base sequence or nucleotide sequence.

The DNA sequence of a gene provides a wealth of information that can be used to probe deeper into the function of its product and the way in which gene expression is regulated. Even for genes whose protein product has not yet been identified and whose function is unknown, a comparison of the DNA sequence with those of known genes and proteins can sometimes provide valuable clues to its identity and the function of its product. In every branch of biology, cloned and sequenced DNAs are being used as a short-cut to the amino acid sequences of proteins and to produce large amounts of a protein in order to study its biological function more easily. Structural motifs within a sequence can be identified representing binding sites within proteins for regulatory molecules such as cyclic AMP, sites for phosphorylation by protein kinases, sites at which carbohydrate side groups may be added, etc.

DNA sequences of the genes for difficult-to-isolate membrane receptor proteins have provided much of our present understanding of their detailed structure and disposition in the membrane. A sequence is also essential before any sophisticated chemical reconstruction can be carried out to produce a DNA tailored for a particular purpose. Worldwide, thousands of bases of DNA sequence-data are being accumulated daily, to add to the millions of base sequence already banked on computer in several centres. Organized data banks, which can be accessed by researchers, have held amino acid sequences and protein structures for many years, and are proving even more necessary for DNA. Two sequencing methods for DNA ate used, developed independently in the mid-1970s by Fred Sanger and his colleagues at Cambridge (the chain termination method) and by Allan Maxam and Walter Gilbert at Harvard (the chemical cleavage method). They both depend on generating from the DNA to be sequenced, a nested set of DNA fragments each terminating before a known nucleotide and which form a series differing by a single nucleotide in length. These can then be separated according to size by gel electrophoresis, and the sequence read off the gel.

RNA genomes can also be sequenced after preliminary transcription into DNA by the enzyme reverse transcriptase. In this way RNA turnout viruses have been sequenced as well as other important RNA viruses such as polio, influenza and the AIDS virus - HIV. DNA sequences have provided many surprises. Notable are the overlapping genes of ΦXM74 and departures from the 'universal' genetic code in mitochondria and protozoa.

The carrier DNAs into which 'foreign' DNAs are spliced to make a recombinant DNA for introducing into a new host cell are often known as vectors. The first vectors used were small self-replicating circles of plasmid DNA from the bacterium *Escherichia coli.* Plasmids also occur in many other bacteria] species. Naturally occurring plasmids typically carry a handful of genes that, for example, specify resistance to antibiotics, the production of proteins (colicins) lethal to other bacteria] species, or useful adaptive enzymes.

Plasmids are typically carried in one or more copies per bacterial cell. Plasmid DNA is easily isolated and separated from the bacterial chromosomal DNA by virtue of its small size. A piece of DNA (up to a certain size limit) spliced into plasmid DNA *in vitro* and reintroduced into a bacterial cell is then multiplied along with the plasmid DNA as it replicates. Some plasmids reach very large numbers of copies per cell, which together with the rapid division of bacterial cells—once every 30 minutes or so—soon produces a large number of copies of the required DNA.

The plasmid DNA is then reisolated from the bacterial culture and the 'foreign' DNA recovered by various means. Bacterial viruses—in particular phage lambda, a virus that infects *Escherichia coli*—have also been developed as vectors. Each recombinant phage infecting a single bacterial cell produces many thousands of progeny, each bearing the 'foreign' DNA. The only parts of a plasmid or phage that are required in its role as a cloning vector are its origin of replication, a specific signal encoded in the DNA that is recognized by the host cell's replication machinery, and for plasmid-derived vectors one or two 'marker' genes (these are usually antibiotic resistance genes) by which cells that contain the recombinant DNA can be easily selected from those that do not. Parts of plasmids and phages that allow them to spread from one bacterium to another are removed in the interests of safety—for example, to prevent any escaped plasmid from infecting the normal bacterial inhabitants of the human body. From the original plasmid and bacteriophage vectors large numbers of purpose-built carrier DNAs have been derived for a wide variety of applications. Vectors derived from plant bacterial plasmids (the Ti and Ri plasmids of *Agrobacterium* species) have extended recombinant DNA techniques to plants to the benefit of plant molecular biology which is now rapidly catching up with its more advanced animal and bacterial counterparts.

The most commonly used bacterial host for gene transfer, both for commercial application and for research, is *Escherichia coli.* It is a natural inhabitant of the human gut and has been for years the workhorse of molecular genetics. The strains used in recombinant DNA work are modified so that they cannot survive outside the laboratory. Other bacteria have also been developed as recombinant DNA hosts but not to the same extent. Species of Bacillus are one potential target as in their unengineered form they are already widely used in the large-scale production of enzymes and fine chemicals (such as the proteolytic enzymes in biological washing powders). Genes can also now be introduced into other bacteria of agricultural interest such as the nitrogen-fixing rhizobia and species of soil Pseudomonas, and into *Agrobacterium* for subsequent transfer to plant cells. Antibiotic-producing streptomycetes, which despite their superficial resemblance to fungi are in fact prokaryotes, are another potential target for genetic engineering.

No bacterial cell will however express an 'unprocessed' plant or animal gene as it does not recognize many of the DNA signals by which eukaryotic gene expression is activated. It is now usual to hitch a cDNA copy of the corresponding mRNA, trimmed to contain only the protein-coding sequence, to an efficient host cell promoter. Vectors that have been constructed to contain suitable promoters and other regulatory sequences in the correct relationship to inserted 'foreign' DNA are generally known as expression vectors. Expression vectors have also been developed for use with yeast and mammalian host cells.

Genetically engineered bacteria cannot of course be used to study the control of eukaryotic genes and neither are they ideal for producing some mammalian proteins. The unicellular eukaryotic microorganism *Saccharomyces cerevisiae* (baker's yeast) is now increasingly used in studies of eukaryotic gene expression, and commercially to produce recombinant proteins. Vectors for use in yeast are derived from yeast plasmids. Shuttle vectors are vectors specially constructed from bacterial and yeast vector components, that enable the genes they contain to be initially cloned in *E. coli* and then directly transferred to yeast cells for expression studies.

Long DNA fragments of 500 000 base pairs or more, much larger than those that can be accepted by plasmid or phage vectors, can also now be cloned in yeast cells, by incorporation into an 'artificial chromosome' (YAC or yeast artificial chromosome) containing a yeast replication origin and yeast telomeres (DNA that seals the ends of the chromosomes and which is needed for successful replication), Vectors derived from animal viruses, chiefly the DNA virus SV40 (simian virus 40), retroviruses and, for some vaccine applications, vaccinia virus, provide replication origins and regulatory signals that allow cloning and expression in animal cells.

The genetic modification of animal and plant cells lagged behind that of bacteria, the chief obstacle being that eukaryotic cells do not take up added DNA as readily as do bacteria.

Animal cells

Various ways of introducing DNA into animal cells and making it more likely that it will be incorporated permanently into the chromosomes and correctly expressed have been developed over the past decade. In suitable conditions, cultured animal cells can now be induced to take up added DNA, albeit at a fairly low frequency. Although the process is analogous to that of transformation in bacteria, where animal cells are involved it is usually called transfection, to avoid confusion with the existing use of 'transformation' to denote the changes seen in cultured mammalian cells when they convert to a cancerous growth pattern. Reliable techniques for transfecting animal cells have only been devised within the past 10 years. The entry of DNA fragments into the cell is aided by placing the cells in a strong electric field (electroporation) or by using a calcium precipitate of DNA. DNA can also be introduced by direct microinjection into the nucleus with a fine glass micropipette. In order to be stably maintained and be expressed in a eukaryotic cell, introduced DNA must not only cross the cell membrane but also enter the nucleus. Injection into the nucleus therefore increases the chances that the introduced gene will become permanently established in the chromosomes. Whole chromosomes or parts of chromosomes can also be introduced into animal cells by fusing two cells to make a somatic cell hybrid.

A way of increasing the chances that the introduced genes will become part of the chromosomes is to splice them into vectors that insert into the chromosomes. These vectors are derived from retroviruses and some DNA tumour viruses such as SV40. Retroviruses are unusual RNA viruses that naturally integrate DNA copies of themselves into the chromosomes where they are passed on from cell to cell and inherited just like cellular genes. They contain sequences (the so-called long terminal repeats (LTRs)) that promote their integration at random into chromosomal DNA. Parts of the viral DNA responsible for integration can be spliced to a 'foreign' gene to make a recombinant DNA that is more likely to become permanently established in the chromosome than the gene on its own. Unwanted bits of the viral genome—for example those that direct the production of infectious virus particles—are deleted. The modified virus DNA is introduced into cells by transfection or microinjection. Ingenious strategies based on retroviral vectors now hold out the best hope of making gene introduction into human somatic cells efficient and reliable enough to be able to correct some genetic defects.

Plant cells

Plant cells are normally encased in a thick wall which prevents many of the procedures for introducing new genes, such as transfection or cell fusion, that are possible with animal cells. But if the wall is removed by gentle treatment, the naked protoplast can be kept alive in culture for a short time. Protoplasts can take up DNA, can fuse with each other to form hybrid cells, and will eventually regrow the cell wall and divide to form an undifferentiated mass of callus tissue. In some plant species at least, a complete new plant can be regenerated from this callus tissue in culture after treatment with plant growth hormones. Protoplasts from different species, or even different genera, can be fused, and this is a promising approach to introducing genes (for disease resistance, drought tolerance, improved nutritional value of plant proteins, etc.) from one species into another.

Agrobacterium and its plasmids

The bacterium *Agrobacterium tumefaciens* infects many 'broad-leaved' or dicotyledonous plants causing tumorous galls ('crown gall') on leaves and stems. This bacterium contains a plasmid—the Ti plasmid—that carries genes required for the transformation and continued proliferation of infected cells. During infection, *A. turnefaciens* transfers a portion of this plasmid—the T-DNA—into the cell's chromosomes, where it becomes stably integrated. This transfer of DNA from bacterium to plant makes *Agrobacterium* a natural carrier for introducing foreign genes. Any DNA spliced into the T region of the Ti plasmid is automatically integrated into the host plant genome. A decade of genetic engineering of the Ti plasmid has produced tailor-made vectors that still transfer their DNA but do not, for example, cause tumours. Many genes have been introduced into whole plants and plant cells in culture either by infection with the bacterium itself or by introduction of the plasmid DNA alone. However, agrobacteria naturally infect only dicotyledonous plants (e.g. tobacco and tomatoes and other vegetables), and the important monocotyledonous crop plants (e.g. cereals and maize) have not been particularly susceptible up to now to genetic engineering with Ti plasmids. Ways of introducing T-DNA into monocots are now being sought, and some success is being achieved. The host specificity of agrobacterial infection seems to be due in part to chemicals produced by the plant itself when wounded—agrobacteria generally enter through wounds. The wound exudate apparently activates a bacterial gene necessary for infection and agrobacteria have, for example, been induced to infect a monocot (e.g. the yam, Dioscorea bulbifera) and produce a gall by treating the bacteria beforehand with wound exudates. Plasmids similar to Ti plasmids occur in another *Agrobacterium* species, A. *rhizogenes,* which causes a proliferation of fine rootlets ('hairy root disease') on the plants it infects. Like the gall tissue produced by A. *tumefaciens,* 'hairy root' tissue also contains DNA derived from a bacteria plasmid—the Ri plasmid—that has become incorporated into the plant's chromosomes. Ri plasmid DNA is also used as a carrier for introducing foreign genes into dicotyledonous plants.

Many proteins of medical and commercial importance are now produce from genetically engineered bacterial, yeast and cultured mammalian cells into which the required gene has been introduced, rather than by expensive and laborious purification from animal or human cadaver tissues in short supply. For proteins used therapeutically—such as insulin, growth hormone and the blood-clotting factors—this enables the human protein rather than an animal substitute to be used, and also eliminates the risk of virus transmission from the original tissue along with the extracted protein. In the past, this has led to contamination of growth hormone with a slow-acting virus that eventually causes neurological damage, and to the anti-haemophilia factor VIII with the human immuno-deficiency virus (HIV) that causes AIDS. It is also virtually the only way of producing large quantities of proteins that occur normally only in minute amounts in tissues.

The first 'recombinant' mammalian proteins were produced in bacteria. But although bacteria grow rapidly, and potentially can produce enormous quantities of protein, they have several drawbacks when it comes to producing mammalian and other eukaryotic proteins. One is that all secreted mammalian proteins, amongst which are most of medical interest, are glycoproteins—that is, they carry carbohydrate side chains attached at specific sites on the protein. The sugar side chains are often essential for full biological activity of the protein. In the eukaryotic cell, the addition of these sugar side chains (glycosylation) occurs in the endoplasmic reticulum. Bacteria do not possess an endoplasmic reticulum and although they have a mechanism for glycosylating their own proteins often do not glycosylate the mammalian proteins that they are synthesizing from transplanted genes. Neither will bacteria secrete mammalian proteins without the addition of a bacterial 'signal' sequence, which has later to be removed. This has led to the search for alternatives to bacteria for recombinant protein production.

Much effort is now going into developing genetically engineered yeast, cultured mammalian cells and even simple organisms such as silkworms, to mass-produce useful mammalian proteins. Several recombinant proteins including insulin and interferon are already produced commercially from mammalian cells, and an: anti-hepatitis vaccine is now produced from hepatitis antigens produced in yeast cells. For commercial production, the transplanted—gene is never in its original state. Various modifications are made to allow it to be transcribed and translated as efficiently as possible in its prospective host cell. All that is required from the original source is the protein-coding sequence; all the other DNA sequences needed to control its expression are supplied on the carrier DNA into which the gene is spliced. Genes intended for bacterial cells are hitched up to an efficient bacterial promoter, those intended for mammalian cell cultures to appropriate cellular (or animal virus) promoters and enhancers. The final recombinant 'gene' is usually a mosaic of elements from different sources. This mosaic is constructed by successive modifications *in vitro* to the vector DNAs. Many 'ready-made' vectors containing the requisite control elements, into which a cDNA or synthetic gene can be slotted as required, are now in use.

The deliberate improvement of the genetic makeup of crop plants and animals has been going on for thousands of years through selection by farmers and, during the past 100 years, by increasingly sophisticated breeding programmes. The development of recombinant DNA technology, however, with its potential for directly altering the genetic makeup of plants and animals in ways not previously possible, looks set in the future to revolutionize plant -and animal breeding. The techniques of gene transfer pioneered in bacteria have over the past decade been extended to plants and some animals—mammals and a few other species.

Many strains of transgenic plants and animals bearing genes from other species now exist. The particular developmental programme of mammals ends itself to *in vitro* fertilization and embryo replacement in ways not immediately possible with other animals. This section deals entirely with transgenic mammals. Genetic modification in other types of animal has hardly been explored except in the case of the fruitfly *Drosophila* where rather different techniques are used to create particular genotypes for genetic and developmental studies. Transgenic plants that produce their own insecticides from transplanted bacterial genes are only one example of the new generation of crop plants that are being developed by a combination of recombinant

DNA techniques and the longer-established methods of plant cell culture. An important limitation in conventional plant breeding programmes is that plants of different species do not naturally interbreed.

Many modern crop varieties are of a different species from their wild relatives, or have become so divorced from them over thousands of years of selection that normal interbreeding is not possible. The wild relatives of crop plants, however, constitute a vast reservoir of potentially useful genes for disease resistance, lerance to cold or drought, nutritional quality, etc. Although plant breeders have devised several ingenious ways of crossing the species barrier, a breeding programme using standard techniques can take years to introduce a desirable trait from a wild plant into a cultivated relative.

The problems usually do not lie in any incompatibility in the genes of the two species at the level of their DNA. DNA is a most accommodating molecule and, as recombinant bacteria show, living cells can accept genes directly transplanted from any source and, with a little extra genetic engineering, be persuaded to produce the proteins they specify. The new genetic techniques can now be used to introduce useful genes from plants and bacteria directly into plant cells, from which, by the well-tried methods of plant cell culture, new transgenic plants can be regenerated.

The current acquisition of seed companies and plant breeding companies by the giants of the agrichemical industry reflects the potential ability of genetic engineers in the long-term to produce seeds with their own inbuilt resistance to fungal and insect pests, and to make more economic use of fertilizers (or even, a more remote possibility at present, fix their own nitrogen). Such plants would not need the current heavy applications of pesticides and fertilizers. Of more immediate value, DNA technology has also provided genetic probes for valuable genes that can be used to screen large numbers of seedlings or cultured cells in more conventional breeding programmes, so that it is not necessary to wait months before finding out, through laborious testing of mature plants, whether the desired genes have been passed on in the genetic lottery.

In animals, the techniques of gene transfer are most advanced in the laboratory mouse. Work with larger, less prolific domestic animals has progressed more slowly but strains of transgenic pigs, sheep, goats and cows have been established.

When advances in reproductive technology that include *in vitro* fertilization, storage of frozen embryos and surrogate motherhood are combined with gene transfer directly into fertilized eggs and very early embryos, a powerful new genetic technology has been created whose potential is only just beginning to be explored. Genetic manipulation in animals aims to incorporate the new gene into the germline so that it can be passed on from generation to generation just like any other gene, and a strain of stable transgenic animals tan be bred. Such manipulation of the human germline—for example, to correct a genetic defect—may be, possible in the long-term, but is not in prospect, for both ethical and technical reasons. However, there are some inherited genetic diseases affecting humans whose effects may be alleviated by transfer of an unaffected gene into somatic cells after birth. Although there are considerable technical problems to be solved, the first trials of somatic gene therapy in humans are undoubtedly not far away.

Transgenic animals are reared from embryos into which new DNA has been introduced by genetic manipulation of a fertilized egg or very early embryo outside the mother. If the new DNA becomes' stably incorporated into the chromosomes at this early stage, it will be present in a large number, if not all, the cells of the developing embryo, including the reproductive cells (germline).

There are various routes for introducing DNA into a fertilized mammalian egg or very early embryo *in vitro*. (In this connection, the mammalian early embryo, sometimes called the pre-embryo, refers to the as yet totally undifferentiated ball of cells that develops within a few days of fertilization.) Up to the blastocyst stage (at which the mass of dividing cells has formed a hollow ball ready to implant into the wall of the uterus) the embryo can be reintroduced into the womb. DNA can be injected straight into the nucleus of the fertilized egg (microinjection) or alternatively into the centre of the blastocyst where it is taken up by at least some of the cells and becomes incorporated into their chromosomes. Another technique is to modify compatible, undifferentiated embryonic stem cells by transfection or retrovirus-carrier infection, and then replace them in the embryo (which can be frozen to arrest its development and stored until needed). Mammalian pre-embryos at this stage are remarkably robust to interference. The cells that will give rise to the embryo itself have not yet been set aside. Unlike the embryos of other species they still retain the potential to replace removed cells or incorporate additional ones without affecting subsequent development. So far, these techniques have been applied to laboratory and some domestic animals, chiefly the laboratory mouse. Work on farm animals has produced transgenic pigs, sheep, goats and cows, but as yet only on an experimental basis. One goal for animal breeders is to be able to produce an animal with required characteristics—such as better reproductive performance or disease resistance—without the present lengthy breeding process. There are still considerable problems to be overcome before this goal might be achieved. These are largely being investigated in the transgenic mouse, which as well as being a testing ground for techniques destined for use in farm animals and human gene therapy is also of interest in its own right in fundamental studies on the control of gene expression.

The most immediate problem is that it is not yet possible to predict where in a genome the introduced DNA will settle. If it disrupts resident gene function, the modified embryo may abort or be born with congenital malformations or other defects. Given that DNA can be introduced without interfering with the expression of the remainder of the genome, there is still the question of whether the introduced genes function effectively, that is, whether they are transcribed into RNA and translated only at the proper times during embryonic and adult life.

At present, the absolute success rate per treated egg or embryo is low. Only a very small proportion of treated embryos end up expressing an Introduced gene. For example, in one recent experiment, of around 390 mouse embryos treated initially, 240 were finally transferred to surrogate mothers to give 45 live offspring, of which 17 carried the transgene. It was only expressed in four of these. When it works, however, the effects of an introduced gene can be dramatic. A strain of transgenic mice carrying a rat growth hormone gene grow to nearly twice the size of normal mice. But, for an individual animal, it is at present impossible to predict whether an introduced gene will be expressed, or whether it will be regulated correctly. In higher organisms, there is at present no completely reliable way of targeting the new gene to a particular place on the chromosome. Experiments with animals and with mammalian cells in culture show that introduced genes can integrate virtually anywhere in the chromosomes, and that the position of a gene on the chromosome can greatly affect its activity and its capacity to be properly regulated, If it comes within the orbit of another gene regulatory system it may be permanently inactivated or become active in the wrong tissues.

However there are numerous cases where genes appear to have become permanently incorporated and are being correctly expressed and regulated. There are ways of making it more likely that the gene will be expressed in the required tissues and at the appropriate time. This involves the use of a 'fusion gene' as the introduced material. The host promoter should respond to the biochemical cues it normally responds to and switch on its associated protein-coding sequences, which may be derived from any source, bacterial or eukaryotic, in the required fashion. This type of construct has been used recently to obtain a transgenic mouse strain that secretes sheep milk proteins in its milk, as a preliminary test of the feasibility of modifying the protein content of cow's milk. A more general problem in the application of direct gene transfer to farm animals is that most of the characteristics one might wish to modify, such as size, reproductive characteristics, growth rate, etc. are the product of many genes acting independently. The individual genes contributing to such polygenic traits are in most cases not known, and it is extremely difficult, if not impossible, to disentangle their individual effects. They may well exist as blocks of co-adapted genes, where a radical alteration in one gene often upsets the, function of the block of genes as a whole. Such traits are therefore not easily amenable to improvement by artificial genetic manipulation.

A start has been made however, by introducing extra copies of their own growth hormone genes into pigs in the hope of producing a strain of pigs that will grow faster without the use of the steroid hormones and antibiotics that are used commercially at present to promote growth. Another idea is that of engineering the genome so that a cow, for example, will produce desirable 'foreign' proteins in its milk. This approach could be used either to improve the protein quality of milk for consumption, or as another way of producing large amounts of commercially valuable proteins that could be easily isolated from milk.

One problem in transgenic work is that the introduced DNA may actually cause mutations by inserting into host genes and disrupting their action. Although this is a grave disadvantage in work aimed at gene therapy it has been exploited by geneticists as a rapid way of producing new mutants that might be useful in mapping large genomes and in developmental research.

Mammalian developmental research generally is severely hampered by a dearth of informative mutations. An advantage of mutations caused by transgenes over those identified purely genetically is that the gene affected by the mutation can be easily recovered using the transgene as a probe, and then used to isolate the normal gene from a DNA library for sequencing, gene mapping, etc. The ability to introduce new genes into i mammalian cells raises the possibility of being able to correct genetic defects in humans by introducing a copy of the normal gene into the appropriate cells.

The techniques are at present not sufficiently reliable to be applied to *in vitro* fertilized early human embryos to correct genetic defects, even if such procedures were considered ethically acceptable or even necessary, given the prospective ability to detect genetically defective eggs and implant only those not carrying the defective gene.

Proposed legislation would ban any experimental genetic manipulation of fertilized eggs in the UK, and similar legislation has been proposed in most other countries. The first targets for gene therapy in humans are tissues such as the easily extracted, self-renewing and transplantable cells of the bone marrow that produce all the cells of the blood and the lymphocytes of the immune system. Genetically corrected bone marrow might also be able to alleviate the symptoms of other conditions that arise from a general lack of a particular enzyme. Some inheritable defects in bone marrow cells are in principle already treatable by a transplant of healthy marrow from a matched donor, but the problems of tissue matching mean that it cannot be offered to everyone who might benefit from it. A year or so ago it was widely predicted that bone marrow replacement would soon be ready for clinical trial in humans. However, there are technical problems of getting a sufficient number of the treated cells to take up the gene and express it correctly, so that at the time of writing, human gene therapy is still untried.

The goal in gene therapy is ideally a clean swap of the good gene for the defective resident genes or at the least to target the incoming gene to a predetermined site on the chromosome where it will be correctly regulated and not interfere with the expression of other genes. Many of the genetic diseases that might be alleviated by corrected bone marrow are caused by defective genes that either make no product at all or an ineffective product. In most of these cases only one copy of a good gene need be introduced and the defective gene need not be removed. Straightforward replacement of a resident gene by its introduced counterpart occurs routinely in bacteria and yeast. Unfortunately, although a similar process can occur in mammalian cells, it is for practical purposes overshadowed by the far greater likelihood that the introduced gene will become incorporated at random in the chromosomes. Although genes integrated at random often appear to function and be regulated normally, in many instances they do not, and the unpredictability of the results presents problems for gene replacement in humans. Nothing is known about the long-term effects over a human lifetime of such an anomalously positioned gene.

The self-renewing stem cells that have to be corrected if the marrow is to continue to generate corrected cells throughout the individual's lifetime have not yet been isolated with any certainty, and are calculated to make up only a tiny proportion of the cells in a sample of bone marrow. If it becomes possible to obtain enriched cultures of stem cells the prospects for reliable gene replacement in bone marrow could look brighter. Vectors likely to provide the required high rates of gene introduction are those derived from retroviruses.

Human bone marrow cells can now be successfully infected with specially engineered retroviruses carrying an appropriate gene, which, when once inside the bone marrow cells, insert a DNA copy of themselves and their added gene into the chromosomes but are unable to produce infectious virus. The first attempts at gene therapy via corrected bone marrow are predicted to be aimed at adenosine deaminase (ADA) deficiency, which produces severe combined immunodeficiency usually leading to early death in childhood, and Lesch-Nyhan disease, a rare X-linked condition caused by a deficiency of the enzyme HGPRT (hypoxanthine:guanosine phosphoribosylt-ransferase) and leads to a bizarre syndrome of uncoordinated involuntary movements and aggressive self-mutilation. Although the effects of HGPRT deficiency are most marked in the brain, it is possible, but by no means certain that a circulating source of HGPRT might alleviate symptoms. Both these enzymes are produced constitutively in most cells and do not present any special problems of gene regulation. Both these conditions are mercifully rare. Much more common are the hereditary anaemias caused by defective globin genes. Until recently however, these most obvious candidates for genetic correction have proved somewhat intractable. The globin genes represent a very tightly-regulated gene system: two separate globin genes have to be expressed to make haemoglobin, they are only expressed in the precursors of red blood cells, and it is not yet known exactly how globin genes are regulated.

Unlike cells and tissues from an adult animal, many plant tissues retain their capacity to regenerate a complete new plant throughout the life of the plant. Cell and tissue culture as a means of plant propagation is widely used commercially, and it is now possible to introduce novel genes into cultured plant cells and tissues. Plants bearing new genes introduced from other plant species and from bacteria have been raised from individual plant cells and plant tissue genetically modified in culture.

The practical application of these techniques to important crop plants still awaits solutions to several problems. Finding suitable conditions for regenerating entire plants from cultured cells and tissues is still largely a matter of trial and error for different species. The staple cereal crops of wheat, rice and maize are monocotyledons and have proved much more difficult than dicotyledonous plants to grow and modify in culture. Tobacco, tomatoes, petunias and potatoes at present figure largely in accounts of genetically engineered plants as they are easy to manipulate with *Agrobacterium* vectors and to regenerate from tissue culture. The first successful transfers of foreign DNA into cultured cereal cells, by transfection of protoplasts, were only made very recently and at the time of writing it is not yet possible reliably to regenerate transgenic cereal plants from a single genetically modified cell, although it is undoubtedly only a matter of time before this will be achieved. One reason that monocots have lagged behind dicots in the genetic-engineering business is that they are not naturally susceptible to infection with *Agrobacterium*, the best-developed carrier of foreign DNA for plant genetic manipulation. Ways of introducing DNA into cereal cells using novel routes and other carriers are now being developed. As well as introducing new genes, it is now also possible to engineer the genome so that some of the plant's own genes are permanently switched off, or their expression changed in some way.

The increasing cost and environmental problems caused by heavy applications of fungicides and pesticides makes built-in resistance against insect pests and the fungal and viral pathogens that afflict crop plants a major aim of any plant-breeding programme. Plants have many ways of resisting attack.

Many possess physical defences against insect damage in the form of leaf hairs and spikes, and others produce noxious chemicals that deter herbivorous insects. They can produce general chemical responses that limit fungal infection, and in some cases genes responsible for resistance against specific fungal pathogens have been identified.

In most cases, natural pest and disease resistance is the product of many genes working together and so are not immediately amenable to genetic engineering techniques. Even where single-gene resistances have been identified, fungal pathogens, with their capacity for rapid variation, soon evolve to overcome them which is why the lifetime of most modern disease-resistant varieties is usually short. For these reasons the conventional methods of plant breeding, backed up by protoplast fusion and tissue culture techniques, are unlikely to be superseded immediately.

Recombinant DNA, however, offers the possibility of attempting some novel strategies and a few examples are outlined below. One important consideration in introducing totally new genes into crop plants or in persuading them to make much larger amounts of some proteins than normal, is that this does not make the plant unpalatable or harmful to eat. As well as altering the plants themselves, much effort is focused on altering the genetic makeup of the bacterial inhabitants of the plant and its immediate environment, and in improving the effectiveness of natural pathogens of the pests.

Strategies being developed by genetic engineers to protect crops against attack by insect pests range from incorporating genes for natural 'insecticides' into the plant to producing more effective forms of the viruses and other microorganisms that attack the pests. Several types of pest-resistance genes have already been transferred experimentally. One is a bacterial gene encoding an insecticidal toxin produced by the common soil bacterium *Bacillus thuringiensis*. The toxins produced by B. *thuringiensis,* which are lethal to the caterpillar stage of many insect pests, are already in use as natural insecticides. They are proteins and are processed to toxic form only in the insect's gut. From the point of view of safety to humans, animals and beneficial insects they are the exemplar of a safe insecticide. The bacteria produce large crystals of toxin, which have been produced commercially and used as insecticides for many years, especially in the production of organically grown crops.

Plants have been raised recently that incorporate the bacteria gene specifying the toxin. They produce the protein in their tissues with no apparent effects on their own growth, and in experimental conditions show considerable pest resistance. However, the toxin is sensitive to ultraviolet light and is soon inactivated. In an alternative approach to the problem, the genes for B.

thuringiensis toxin have also been transferred to bacteria of the genus Pseudomonas, common inhabitants on the leaves of many plants, in the hope that live Pseudomonas producing the toxin may also act as a deterrent. Another approach to insect resistance is to incorporate genes for proteinase inhibitors. These are small proteins occurring naturally in some plants which, when taken in by caterpillars, interfere with the activity of digestive proteinase enzymes such as trypsin. The caterpillar is unable to digest and make use of the plant protein and starves to death or fails to develop. A gene for the trypsin inhibitor of cowpea (Vigna unguiculata) has recently been introduced experimentally into tobacco (Nicotiana), where it significantly deters tobacco's own herbivorous pests, such as the tobacco budworm. (This particular trypsin inhibitor appears to have no effect on human trypsin.)

Another avenue to protecting plants against insect attack is to exploit the viruses and bacteria that naturally infect the insects. This approach is being investigated in the UK at present to control the caterpillars of the moth *Panolis flammea,* a pest of the introduced lodgepole pine (Pinus contorta), now planted in large numbers in Scotland. In some parts of Britain the caterpillars are subject to natural control by a baculovirus parasite. The virus, however, is absent from the Scottish plantations. Introduction of the natural baculovirus as a biological control agent has been under trial for the past 2 years. It might be made more effective, however, if engineered to contain genes specifying, for example, a caterpillar-killing toxin. This was the first release of a genetically engineered organism into the environment in the UK. The aim eventually is to produce a virus that is more effective against its insect pest,, but which retains its natural strict host—specificity and inability to persist in the absence of the host.

Virus diseases pose a considerable problem to agriculture and horticulture as they cannot be controlled by chemicals. Viruses spread by sap-sucking or leaf-eating insects can to some extent be prevented by controlling the insect population or by growing stocks for propagation in areas free of the insect carrier. Virus-free stocks of many, ornamentals, soft fruit, etc. are now also produced by propagation from tissue cultures of virus-free shoot tips. Natural resistance to viruses also occurs and can sometimes be incorporated by cross-breeding. Several new molecular genetic approaches to virus resistance involve a detailed knowledge of virus molecular biology and the use of recombinant DNA techniques. One approach depends on incorporating the gene for the virus's own coat protein into the plant's DNA. A viral particle consists of a core of RNA or DNA surrounded by a protective protein coat, often composed of only one type of protein—most common plant viruses are RNA viruses. Viral RNA entering the cell on infection is then immediately mopped up by the coat proteins produced by the cell, preventing its multiplication. Coat protein genes for tobacco mosaic virus (in this case, DNA copies of the virus RNA) have proved a partial protection at least to tobacco mosaic infection.

Another strategy is directed against those viruses that are sometime accompanied by 'satellites'—small circular RNAs that are not part of the virus and which appear to lessen the severity of infection. The satellites are of considerable biochemical interest in their own right as they belong to the diverse group of 'ribozymes'- RNAs with enzymatic activity. DNA copies of the satellite RNA incorporated into the plant genome are activated to produce large amounts of satellite RNA on infection by their corresponding virus and have been shown to provide protection against, for example, the tobacco ringspot virus.

Engineering crops like cereals to fix their own nitrogen at present remains a long-term aim. The natural nitrogen-fixing symbiosis between rhizobia (the nitrogen-fixing bacteria) and the roots of leguminous plants depends on the action of many plant as well as rhizobial genes to establish the symbiosis and accommodate the complex biochemistry involved in nitrogen fixation by the bacteria] enzymes. Conventional genetics has, however, been deployed for some time in improving the efficiency of nitrogen fixation within the important legume crops (such as soybean) by, for example, extending the host range of particularly efficient rhizobial species, and can now be supplemented by recombinant techniques.

The genes involved in symbiosis can now be analysed more readily thanks to the ability to isolate and clone plant and bacterial genes and to study their products and the factors that regulate their expression. Many of the plant and bacterial genes that are needed to establish symbiosis have now been cloned and their gene products identified. Free-living, soil nitrogen-fixing bacteria, such as Klebsiella *pneumoniae* and Azotobacter *vinelandii*, contribute greatly to the fertility of soils and are also now amenable to genetic modification.

Species of the bacterium Pseudomonas are also targets for agriculturally directed genetic engineering. Varieties of Pseudomonas syringae and other pseudomonads are chiefly harmless inhabitants on the leaves of many plants and in the soil. The best known example of a genetically engineered *Pseudomonas* is the so-called 'ice-minus pseudomonas'. Some forms of *Ps. syringae* have a gene specifying an ice-nucleation protein, which is a component of the bacterial cell membrane and induces ice crystal to form around the bacteria at around—4°C. This causes frost damage to plants at warmer temperatures than if the protein is absent.

(A form of Ps. *syringae* lacking this gene also occurs naturally and makes up a small proportion of the normal population.) The ice-minus pseudomonas is intended to compete with and prevent the ice-forming Pseudomonas strains colonizing leaves and blossoms.

In the past most of our knowledge of human genetics has come from the study of heritable genetic defects and diseases. Several hundred heritable conditions involving defects in single genes are now recognized, some merely inconvenient to those affected, such as red-green colour blindness, whereas others are fatal in childhood if not treated. Most of the latter are very rare, but a few, such as sickle-cell anaemia, cystic fibrosis and Duchenne muscular dystrophy, are relatively common in certain populations. The single-gene defects are recognized by their strict mendelian inheritance in affected families. Most heritable diseases only show the full-blown symptoms in homozygous individuals—that is, someone who has inherited a defective gene from both parents. Heterozygotes, who have inherited a defective gene from one parent and a normal gene from the other, often show few if any symptoms, although their carrier state can sometimes be detected by biochemical tests; and now by direct analysis of their DNA.

This type of mutant gene is in genetic terminology said to be recessive to the normal allele. Potentially lethal recessive mutations are able to persist in the population because, even though homozygotes usually die before they reach reproductive age, as long as the defective gene remains rare heterozygotes far outnumber homozygotes. There are far fewer dominant inherited conditions, where the disease shows up in individuals carrying only one defective allele. Dominant inherited diseases can persist in the population only if they allow those affected to reach reproductive age and pass the gene on to their children. Huntington's chorea and polycystic kidney disease are two of the best known. The predisposition to some cancers is also inherited as a dominant trait. Huntington's chorea, for which the underlying defect is not yet known, characteristically leads to rapid general mental deterioration from middle age onwards. Recessive conditions for which the underlying defect is already known include many inborn errors of metabolism' caused by a tack of a particular enzyme as a direct result of a mutation in the gene that specifies it. These include phenylketonuria, alcaptonuria, Tay-Sachs disease and adenosine deaminase deficiency. The hereditary anaemias such as sickle-cell anaemia and the thalassaemias are also recessive defects, in this case all caused by structural or regulatory defects in the globin genes that specify the protein components of the oxygen-carrying haemoglobin molecule. One of the commonest genetic diseases in the caucasian population -cystic fibrosis-is also recessive. At the time of writing the basic defect in cystic fibrosis has not yet been finally identified, although there is strong biochemical evidence implicating a defect in the regulation of chloride transport across the membranes of epithelial cells.

The defective CF allele has recently been located and is now in the process of being isolated. If its gene product can be identified, this should provide a clue to the exact nature of the regulatory defect. The conditions mentioned above are due to genes located on the autosomes. Diseases caused by genes carried on the X chromosome show a different and characteristic pattern of inheritance. Recessive X-linked genetic diseases such as haemophilia A (caused by non-production of blood clotting factor VIII), Duchenne and Becker muscular dystrophies and Lesch-Nyhan syndrome (caused by a deficiency of hypoxanthine:guanine ribosylphosphotransferase or HGPRT) are transmitted through symptomless heterozygous female carriers (females carry two X chromosomes) and show up only in males who inherit the defective X chromosome (males carry only one X chromosome, inherited from the mother, and a Y chromosome inherited from the father). In all recessive genetic diseases where the underlying cause is known, the defect results in non-production of the gene product or production of a non-functional product. The heterozygous carriers show few if any symptoms as in these cases one normal gene is sufficient to produce enough of the required protein.

Identifying the gene concerned in some rare heritable conditions may throw light on the causes of much commoner, non-heritable diseases and congenital defects that show very similar symptoms. For nonheritable Alzheimer's disease, atherosclerosis, manic-depressive illness, cleft-palate and some forms of cancer for example, there are rare 'familial' heritable forms which have been traced to a defect in a single gene.

Genetic maps

Gene mapping is one of the oldest exercises in genetics. Long before geneticists had the least idea of what genes consisted of or what they did, they found that certain characters tended to be inherited together, in apparent violation of the mendelian law of independent segregation. Genes were found to fall into distinct linkage groups, each group representing a set of loci that do not behave entirely independently of each other in experimental crosses. A linkage group represents genes carried on the same chromosome. The linear order of the genes within each linkage group can be determined by the degree of recombination that occurs between them when homologous chromosomes are paired at meiosis. The further apart they are, the greater the likelihood of recombination. By counting the proportion of recombinants in the progeny, the distance between the loci in arbitrary map units can be calculated. Linkage groups determined genetically from experimental crosses can be assigned to their corresponding chromosomes by cytogenetic studies. This is easiest in organisms such as Drosophila, where the location of a mutant allele can often be determined by the consequent disruption in the fine banding pattern of the polytene chromosomes. For prolific organisms such as the fruitfly Drosophila, where many gene loci have been identified by mutation, recombination frequencies and thus distance apart in map units can be calculated with some precision.

Recombinational analysis in humans is less easy. Only genes that are naturally present in the population in atleast two different variants (alleles) with easily distinguishable phenotypes can be mapped in this way. Even when a suitable family is found, the true frequency of recombination between pairs of loci is difficult to determine accurately because of the small number of offspring in each generation. Nevertheless, around a hundred loci have been roughly mapped by these methods over the years, mostly for genes encoding enzymes. In some cases data from many different families can be pooled to give statistically reliable results. With the advent, first, of somatic cell genetics, and more recently, recombinant DNA techniques, human gene mapping has been transformed, and detailed maps of the human genome are for the first time becoming a reality.

Gene maps have many uses. In general, a detailed map makes it easier to clone newly identified genes by working along from the nearest known landmark. Much of the impetus for developing a map of the human genome comes from the need to explore links between genetic constitution and disease.

The maps of the human genome now in preparation will make it easier to locate and isolate genetic loci involved in disease, not only for the inherited single-gene defects, but for far-commoner conditions, such as diabetes, heart disease and cancer. These and many other common diseases are believed to have predisposing genetic components, and cancer is now known to involve alterations in specific genes in the cancer cells themselves. Once disease genes have been cloned they can be used as the basis for DNA-based diagnostic tests to detect those at particular risk, and to study the underlying genetic contribution to many common diseases.

Recombinational analysis cannot by itself determine which linkage group corresponds to a particular chromosome. The X and Y chromosomes are the exceptions as some mutations in genes on these chromosomes have an easily recognizable pattern of inheritance. Gene loci can be mapped to a chromosome through mutations which, as well as causing an obvious phenotypic effect cause visible cytogenetic defects such as a change in length and/or banding pattern of the chromosomes that carry them. Once several loci on each chromosome have been established independently in this way, other loci can be assigned by their linkage to these chromosome markers.

In *Drosophila*, the existence of giant polytene chromosomes with their distinctive banding patterns makes cytogenetic analysis a powerful mapping method. In humans (and other mammals) only relatively large chromosomal alterations can be detected and additional ways of assigning genes to chromosomes have had to be devised. These make use of somatic cell hybrids between human and other mammalian cells (usually those of mouse or hamster).

Mammalian somatic cells from two different species can be induced to fuse by a variety of treatments to give a cell with a hybrid nucleus containing both sets of chromosomes. The chromosomes of the two different species can be distinguished from each other by their staining patterns.

During culture, these hybrid cells tend to lose chromosomes of one or the other species. The loss of individual human chromosomes can be correlated with the loss of the cell's capacity to synthesize a particular human protein, thus assigning the gene for that protein to the lost chromosome. Originally this technique could only be used for genes whose gene product is known and can be easily assayed and distinguished from the corresponding animal protein. With the advent of recombinant DNA techniques, however, it is now possible to locate not only known protein-coding genes, but any length of DNA for which a specific probe is available. The positions of genes relative to each other can sometimes be ascertained in somatic cell hybrids by fragmentation mapping. When the hybrid cells are irradiated, breaks in the chromosomes are produced.

The loss of an identifiable fragment, correlated with a lost ability to produce say, a particular enzyme, locates the gene for that enzyme on the lost fragment. In the past few years, as well as phenotypically detectable gene variants, additional markers in the form of restriction length polymorphisms (RFLPs) have been used to provide extra landmarks on genetic maps. More than 800 sites detected by RFLPs and specific nucleic acid probes have now been mapped, and the numbers are increasing rapidly. As the human gene map becomes fuller the easier it is to fit new genes into place.

Any DNA sequence for which a specific nucleic acid probe is available can be mapped by direct *in situ* hybridization of the radioactively or chemically labelled probe to intact chromosomes in specially prepared cells. The label marks out the location at which the probe has bound to its corresponding chromosomal DNA sequence. Much of the progress has been the result of techniques based on recombinant DNA. As more genes are cloned and sequenced, genetic maps are being gradually correlated with physical maps built up from ordered sequences of the DNA fragments cloned in DNA libraries. DNA libraries consist of collections of DNA fragments representing all the DNA in a genome.

There is considerable redundancy and the fragments overlap. They can be arranged in an ordered series by using one fragment as a probe to recognize the overlap and pull out the next, and so on. Each series represents the 'sequence of DNA along an individual chromosome, with fragments containing known genes providing the necessary genetic guideposts along the way. Such maps are at present being prepared for several species including humans. When complete they will be of great value in linking genetic loci whose mutant character and position on the chromosome, but not much else, is known, to their corresponding cloned DNAs. The actual length of a gene in base pairs and in some cases the distance from one gene to the next can then be determined from the physical map, grounding the genetic map firmly in reality. So far, around 400 mapped human genes have been cloned and their sequences determined.

In the absence of any clues as to what a gene may specify, virtually the only way to isolate it is to first map it genetically and then work along the chromosome from the nearest identifiable site using ordered cloned fragments of DNA and sampling the DNA sequence along the way this is called chromosome walking. The initial mapping often depends on chance—for example, the discovery of a patient whose disease is linked to an obvious cytogenetic defect. The recent isolation of the human DMD locus depended on such a lucky chance. The largest piece of DNA that can be cloned in a conventional DNA library using, for example phage lambda as vector, is around 40 000 kilobases. Chromosome walks covering hundreds of thousands of bases therefore assume heroic proportions.

As such distances commonly have to be travelled over to get from a known marker to the site of the 'unknown' gene, new techniques that cover the ground quicker are being devised. Chromosome jumping consists of using specially prepared clones that contain two sites a long distance apart on the chromosome—ingenious tricks are used to delete the DNA in between. It has also recently become possible to clone very long DNA fragments up to 500 000 kilobases in yeast cells. Until recently it was impossible to separate DNA corresponding to individual chromosomes. New methods of electrophoretic separation are now making this possible, and DNA libraries composed of cloned fragments of a single chromosome make the task of assigning genes to particular chromosomes much easier. Ambitious projects to sequence a complete human genome are now in the assessment stage in several countries and will undoubtedly eventually go ahead. Current estimates for the US project put costs at around $200 million annually over the next 15 years for the 3000 million base pairs of the human genome.

When the project was first proposed a few years ago, the timescale was calculated at around a hundred years, but the foreseeable development of fully-automated sequencing techniques now make it more feasible. A complete human DNA sequence would make work on all aspects of human genetics much less technically onerous. It would be an invaluable aid in locating and isolating genes involved in human disease, as well as providing definitive data for basic research into gene organization, expression and evolution.

Complete genome sequences for representative organisms would allow the most exhaustive comparison possible of the changes that take place at the molecular level during evolution. However, the usefulness of such a project, given its cost, has been questioned by some biologists. Large portions of the genome are repetitive And apparently have no coding or gene regulatory function and are unlikely to repay detailed sequencing with much useful information. Regions containing genes of interest are already being rapidly sequenced by individual research groups, and even here much of the information remains uninterpretable, as the underlying biochemistry and physiology on the function of the encoded proteins in vivo have not yet been done.

The ability to create entirely novel combinations of genes from different organisms raised questions of possible health and environmental hazards from genetically engineered microorganisms that might escape from the laboratory.

There was also disquiet about the very idea of manipulating living organisms in this way, and its possible abuse. Fears over practical—risks focused on the use of the bacterium *Escherichia* coli as the host for gene transfer, and whether genetically engineered *E. coli* might colonize the human gut with harmful effects, especially if genes for lethal toxins or genes with the potential to cause tumours had been introduced. The moratorium was observed for almost 2 years within which time regulatory bodies and experimental guidelines were established by governments and major scientific funding agencies worldwide. In the USA influential regulatory bodies are the Recombinant DNA Advisory Committee

(RAC) of the National Institutes of Health, the Environmental Protection Agency (EPA) and the Food and Drug Administration (FDA); in the UK, the Advisory 'Committee for Genetic Manipulation (ACGM) together with the Health and Safety Commission (HSC) is responsible for formulating (voluntary) guidelines, monitoring their observance, and also, in more recent years, overseeing the deliberate release of recombinant organisms to the environment. In the early days of recombinant DNA work, a small number of possible experiments were banned.

The remainder were categorized according to their possible risks as perceived at the time and conditions were laid down under which each type of experiment should be carried out. These conditions specified the required degree of both physical and biological containment. Physical containment refers to the security of the laboratory and the equipment needed to protect the operator and prevent microorganisms escaping to the outside world. Biological containment deals with the microorganism used as host for the recombinant DNA and the DNA carriers (vectors) into which the foreign DNA is spliced. For some experiments specially 'enfeebled' strains of *E.coli* or other bacteria that are quite unable to live outside the laboratory or colonize the human body must be used.

As experience with recombinant DNA has accumulated, and results from the research itself have been able to dispel some of the original fears, guidelines have been relaxed for many categories of experiment. Laboratory research and commercial production of genetically engineered products from microorganisms are generally conducted under the safety conditions one would normally use for dealing with the most hazardous of the organisms involved. The regulatory bodies now have to address the new question of the release of genetically engineered microorganisms into the environment, as for example, pest control agents or live vaccines. Most scientific opinion sees nothing—inherently unsafe about genetically engineered microorganisms and favours dealing with each case on its merits, applying the criteria that a naturally produced biological agent would have to meet before it could be released. For biological pest-control agents, these include non-toxicity to humans and animals, strict host specificity and inability to survive without its host.

Questions of safety aside, the new techniques raise various ethical questions, primarily arising from the ability they bring to alter the genetic make-up of plants, animals, and even humans in ways not previously possible. *In vitro* fertilization and its associated reproductive technology combined with the potential of *in vitro* genetic modification mean that one day it may be possible to produce and raise a human embryo that has been genetically modified. In the UK, the problems raised by the new reproductive techniques and research on human embryos in general were considered by the Warnock Committee.

Among the committee's many recommendations was that some research should be allowed on embryos cultured in the laboratory for up to 14 days after *in vitro* fertilization. At this stage a human embryo is still a tiny ball of cells -the blastocyst—which *in vivo* would have just implanted into the uterine wall. The cells that form the embryo proper have just become differentiated from those that form the extra-embryonic membranes. Proposed legislation in the UK has made no firm recommendation on this particular issue. Members of Parliament are to be given the choice of voting either to ban all such experimentation completely, or to allow a limited range of experiments under licence on cultured embryos up to 14 days after fertilization, subject to the consent of the donors of eggs and sperm. Use beyond this time would be a criminal offence. The limited experimentation allowed would be aimed chiefly at overcoming problems of infertility and improving diagnosis and discovering the origins of congenital defects. Rapid technical advances mean that clear guidance on what is or is not permitted is now urgently needed. Working under current voluntary guidelines, developmental biologists have recently been able to determine the stage at which transcription starts in human embryos using spare fertilized eggs from an *in vitro* fertilization anti-infertility programme. Similar techniques might also be used to determine whether An *in vitro* fertilized embryo from a mother at risk was carrying a genetic defect. It is not only experimentation on early embryos that poses immediate questions to be resolved. Brain tissue from aborted fetuses has recently been transplanted into the brains of patients with Parkinson's disease, in the hope that the foetal brain cells will supply the dopamine lacking in this condition

Recombinant DNA has also brought in its train the question of patent rights to living organisms. Until recently it was impossible to patent any living organism, even one so highly bred that it would be unlikely to have arisen without deliberate selection. Applicants are required to prove that the animal constitutes a manufacture or composition not found in nature. The first such patent was granted recently to Harvard University, for an 'oncomouse', a strain of transgenic mice carrying an oncogene. Opposition to patenting of animals so far has focused largely on the moral issue, with its implications that humans can create and claim ownership for a new living thing. The question of plant breeders rights and the possibility of being able to patent genetically engineered seeds and the means by which they have been obtained, has taken on a more political dimension. It has sharpened the long-standing grievance felt by Third World countries that although most plant genetic resources are located in the Third World, they are collected and exploited to produce new commercial varieties mainly by the developed countries.

The environment of an organism, the totality of factors that influence its activities, achievements, and ultimate fate, can be difficult to characterize. Many features of organisms are adaptations to their environment. Indeed, mcuh of biology, whether it be biochemistry, anatomy, physiology, or ecology, consists of the study of adaptations, those features that, having evolve by natural selection.

It is important to bear in mind that not all of evolution consists of the development of adaptations by natural selection; many other factors, including chance, influence evolution. But even if we restrict our attention to evolution by natural selection, there are numerous selective factors besides those imposed by the external ecological world. Chief among these are the internal relationships among biochemical and developmental pathways, and among different organs, that impose selection by requiring that new features be compatible with the rest of the organism's internal organization.

The ecological environment includes both abiotic and biotic features. Climate, salinity, soil type, availability of water, and other physical and chemical features are important. Also, other species—including prey, predators, pathogens, competitors, and mutualists—are a preeminent feature of every species' environment. Moreover, other members of an individual's species, with which the individual may mate, compete for resources or mates, or interact in various social contexts, are also important features of its environment. Thus properties of the population such as its density, sex ratio, or genetic composition may impinge importantly on each individual's prospects of survival and reproduction. The features of the environment that are important vary from species to species because of their different evolutionary histories; it is not an exaggeration to say that by virtue of past evolution, species create their own environments.

To a predatory beetle, the chemical composition of the plants among which it forages is largely irrelevant; but if a beetle evolves the habit of herbivory, plant chemicals that act as toxins, repellents, or feeding stimulants become all-important. Beetle species whose larvae feed on the upper surfaces of leaves contend with a much hotter and drier environment than those that feed on lower surfaces; the beetle larvae that bore in plant stems inhabit a different environment entirely. Not only the past evolution of a species determines its environment; so do its present activities, for species deplete resources, release toxic metabolites, and alter their surroundings in numerous other ways. Thus species and environment alter each other reciprocally. It is an error to think of species simply as passive sufferers of harsh external fate; they are active participants in a dialectical interchange between organism and environment.

Each of a great many factors affects the ability of a species to survive and reproduce, and thus in part determines whether it can persist in a particular locality. For example, a clam might tolerate a certain span of temperatures and feed on plankton in a specific size range. Following Hutchinson, we may draw a two-dimensional graph in which a single point represents an environment with a particular temperature and a particular prey size. Part of this space then represents the range of possible environments in which the species can persist. We may add a third axis corresponding to, say, salinity, thus defining a three-dimensional space containing a region that represents tolerable combinations of temperature, prey size, and salinity.

To take into account other environmental factors we would have to draw many more axes. We conceive of an n-dimensional space, with one axis for each of n environmental factors. Within this space is a region consisting of a cloud of points, each representing a particular combination of temperature, prey size, salinity, copper concentration, starfish abundance, and so forth, that together constitute environments conducive to the survival and reproduction of the clam population. Hutchinson called this region, consisting of the set of possible environments in which the species can persist, the fundamental *Ecological niche* of the species. Because different genotypes may have different tolerance limits for various axes of the niche, the ecological niche can evolve, and indeed will often vary among different geographic populations of a species.

The distribution of a species may be described in geographic terms (e.g., the bog turtle *Clemmys muhlenbergi* is restricted to eastern North America) and in ecological terms. Geographic distributions are often limited partly by history; the species may not yet have dispersed to potentially suitable areas. Some geographic limits are also set by ecological factors such as climate or competition with other species. Biogeography, the study of geographic distributions of species. Within its geographic range, the distribution of every species is patchy to varying degrees because of spatial variation in physical features, the availability of resources, and other species that act as competitors, predators, or parasites. Many species, especially those that have very exacting requirements, consist of several or many small populations that are often widely separated. The recognition that a species is divided into local populations that often consist only of dozens or hundreds of individuals has had an important impact on our understanding of evolutionary processes. Moreover, small populations are very susceptible to extinction, and the persistence of a species in a particular locality often depends on a continual influx of migrants from other populations. For example, what appeared to be a single population of checkerspot butterflies actually proved to be three subpopulations that fluctuated independently in size; over the course of ten years, subpopulations became extinct on several occasions, and were reestablished by new immigrants. Suitable habitats for a species that has a low capacity for dispersal may be unoccupied because they have not been recolonized.

In any locality the density and persistence of the population of a species depends on its capacity to increase in numbers and the factors that limit its abundance. An understanding of these population dynamics is essential to an appreciation of evolutionary theory. In the most common formulation of population growth, we suppose that the rate of change (dN/dt) in the number of individuals in a population (N) is determined by birth (b) and immigration (I), and by death (δ) and emigration (E). For simplicity, immigration and emigration are usually ignored by supposing them equal, but as we have just noted, immigration may be crucial to the persistence of the population.

If b is the per capita birth rate and δ is the probability that an individual will die in the time interval dt, the per capita rate of change in population size is $r = b - \delta$ and the population as a whole grows at the rate

$$dN/dt = bN - \delta N = rN$$

Then N_t, the population size at time t, depends on r and the initial size N_o:

$$N_t = N_0 e^{rt}$$

In populations such as annual insects that have discrete generations, the replacement rate R is used instead of r, and the expression for population growth is $N_{t+1} = RN_t$. R may be defined as the ratio of the number of individuals born in two successive generations.

As long as the birth rate exceeds the death rate ($r > 0$), the population will increase in an exponential fashion; if unchecked, as Darwin noted, the descendants of a single pair of any species would cover the earth in short order. The rate of increase r actually depends on age structure, for a population made up mostly of individuals at the height of their reproductive powers grows more rapidly than if either juvenile or senile individuals predominate. However, if the environment is constant so that birth and death rates specific to each age class do not change, the population ultimately attains a stable age distribution: the entire population grows at the rate r, but the proportion made up of each age class remains constant. As a population grows, the age-specific birth and death rates change, so the actual rate of increase r is not constant. At very low densities r may be less than maximal because survival and reproduction may depend on favourable social interactions or the need to find mates. But as density increases, the population may deplete its resources, poison its environment with metabolic wastes, or engender the build up of predators or pathogens.

Population growth then becomes *Density-dependent*. In the most commonly used (and unrealistically simple) model of density-dependent growth, it is supposed that as density increases, the birth rate declines and/or the death rate increases linearly, so that the actual growth rate r declines with slope c from its intrinsic value r_m according to the equation:

$$r = r_m - cN$$

A little algebra and the relation $r_m/c = K$ yields the *Logistic equation* of population growth in a limited environment:

$$dN/dt = r_m N(K - N)/K$$

According to this equation, population growth proceeds sigmoidally, slowing down and finally stopping when the equilibrium density $N = K$ is reached. Property speaking, K is merely the equilibrium density, which may well be set by, say, predation; but it is often referred to as the *carrying capacity* of the environment. This phrase is suitable if density is limited by scarcity of resources such as food or space.

All species have the capacity to grow exponentially, but none do so indefinitely. In reality the approach to equilibrium is seldom a smooth sigmoid curve, for several reasons. One of these is that the effect of density on birth rates may not be felt until some time after the density has increased (i.e., there is a time lag). When this occurs, the population may oscillate in numbers as it approaches equilibrium and even oscillate violently and unpredictably around its theoretical equilibrium density rather than settling down to a constant size. In most natural populations there is considerable *Reproduction excess*: more individuals are born than survive to reproduce. This realization was critical in Darwin's formulation of the idea of natural selection; as he said in *The Origin of Species*, " . . . as more individuals are produced than can possible survive, there must in every case be a struggle for existence, either one individual with another of the same species, or with the individuals of distinct species, or with the physical conditions of life. . . . owing to this struggle, variations, however slight and from whatever cause proceeding, if they be in any way profitable to the individuals of a species, in their infinitely complex relations to other organic beings and to their physical conditions of life, will tend to the preservation of such individuals, and will generally be inherited by the offspring." Density dependent reduction of population growth is not necessary for the operation of natural selection, but it does offer considerable opportunity for selection to occur.

The factors that limit population density in nature are the subject of a long-standing controversy. Andrewartha and Birch, among others, argued that many populations are in a state of effectively continual exponential growth that is interrupted by environmental changes such as inclement weather before they become dense enough to cause a scarcity of resources or a buildup of natural enemies. Controlling factors such as weather are not responsive to population density, and so are termed density-independent. A contending school of thought denied the importance of density-independent limiting factors, and argued that most populations usually are kept near equilibrium by density-dependent factors such as a shortage of energy, nutrients, nest sites or other resources, predation and disease, or, in some animals, behavioural interactions such as territoriality.

Probably the most common current view is that most populations experience both density-independent and density-dependent growth at different times, with the relative proportions varying among populations and species. There is abundant experimental evidence that plants and sessile intertidal invertebrates are often limited by competition for space (and light and water in the case of plants); that many populations of predators are limited by scarcity of food; and that predators can act as density-dependent limiting factors of their prey, such as phytophagous insects and other herbivores. But it is also the case that weather can cause fluctuations in populations that may well be density-independent.

Density-independent growth may be commonly experienced by *Fugitive species* that occupy very temporary resources and persist only by continual dispersal from one patch of resources to another. The fly larvae that develop in rotting fruits or mushrooms occupy a resource that may become uninhabitable before the density of larvae can become very great. But even in this case, there is evidence that mushroom-feeding flies are food-limited and compete for resources. It seems likely that most populations that persist for very long are at least occasionally limited by density-dependent factors.

Organisms have many responses that mitigate the ill effects of high density. Most notable is the tendency to disperse. Virtually all plants, for example, have adaptations for dispersing their seeds; one of the worst places for a seed to germinate is next to the parent plant, Which is not only a bigger competitor but is also a focus of activity for herbivores and pathogens. In some species of animals, high density triggers a complex of adaptive responses that include dispersal. Many species of aphids (plant lice) that are wingless under favourable conditions give birth to winged offspring if they are crowded or if the quality of their food plant deteriorates, as it does when it is heavily infested. As successive generations of the plague locust *Schistocerca gregaria* experience more and more crowding, these grasshoppers experience a hormonal change that affects their physiology, morphology, and behaviour. They store fat, develop long wings and increased pigmentation, become more gregarious, and finally depart in enormous swarms, to land, with luck, in greener pastures.

Too sparse a population, however, may also be inimical to an individual's reproductive success. Thus fishes find protection from predators by schooling, birds may find food more effectively by watching other members of a flock than by foraging singly, and

Drosophila larvae, by burrowing through their medium, promote the growth of the yeast on which they feed. There are many advantages of aggregation and social behaviour that can outweigh the disadvantages that aggregation necessarily entails, such as competition for food. The limiting factors that prevent unlimited population growth may vary in time and space; for example, a population of herbivorous insects may commonly be controlled by natural enemies, but occasionally escape this control, reach higher density, and become food-limited.

Moreover, several factors may operate simultaneously to govern population density. Birth and death rates may be sensitive to different factors as density increases; thus it is possible, for example, for scarcity of food to lower the birth rate and for increased predation to increase the death rate simultaneously. It is extremely important to distinguish the effects an environmental factor has on a *population* from the effects it has on *individuals* within the population; it is the latter, as we shall see, that determines evolution by natural selection. Thus a population may be limited by food rather than predation, but some individuals nevertheless suffer predation, and there will be selection in favour of genotypes that can avoid predators. Predation may be "good" for a prey population by preventing it from becoming extinct by overexploiting its resources, but predation is not good for the individuals that suffer it.

The subject of density-dependent population growth is integrally related to that of interactions between predators and their prey, or, more generally, between consumers and the species they consume. Because a consumer feeds on a dynamically growing population, its carrying capacity is not a constant, but a variable. As the prey population increases, that of the predator does likewise until it becomes so abundant that the prey population declines. The two populations, then, mutually limit each other's density: the prey population is limited by predation, and the predator population by food. In theory, either stable or unstable oscillations in the densities of both species may occur; if unstable, extinction of the prey, and thereafter the predator, may result. In many cases, predator-prey interactions are indeed unstable, and result in extinction.

For example, a South American cichlid (*Cichla ocellaris*) introduced into Gatun Lake in Panama has locally extinguished several other species of fishes. The interaction may be more stable, however, if the prey species itself is partly limited by the availability of its own resources, so that it does not grow as rapidly when predation is reduced. Although some predator-prey combinations are unstable, other combinations persist for any of several reasons. Predators are often unable to find all the potential prey, some of which remain inaccessible.

A prey species may survive through dispersal: the interaction may be locally unstable, so that local populations become extinct, but populations of prey, and consequently of predators, may build up elsewhere as vacant localities are colonized. In some instances, some of the age classes in a prey population are resistant to predators; for example, wolves attack chiefly defenseless young and infirm old moose rather than the reproductive age classes, which can effectively defend themselves. Moreover, many predators (and other consumers) consume several species. In some cases, predators focus their food-seeking efforts to a disproportionate degree on the most common kinds of prey. Whether by moving to places where a particular food type is concentrated, or by forming a *Search image* that enables them to find food more effectively, they lessen their impact on uncommon species of prey.

Distasteful animals frequently advertise their unpalatability by a warning (*Aposematic*) coloration or pattern; after a few encounters with such distasteful organisms, naive predators quickly associate the colour pattern with their unpleasant experience and refrain from attacking such prey for some time.

Aposematically coloured species are often the models in systems of mimicry for convergence to a common colour pattern in otherwise dissimilar species. In some cases a palatable species masquerades as an unpalatable one (*Batesian mimicry*), while in other cases several unpalatable species converge in appearance, each species gaining protection from its similarity to the other ones (*Mullerian mimicry*). A community of organisms displays a bewildering variety of defense mechanisms, which continue to evolve in diversity. If two species share the same defense system, the population of a predator adapted to counter this defense grows as a function of the combined abundance of the two species of prey. One prey species may therefore become extinct, because the other maintains the predator at high density. Thus the dynamics of two such prey populations, viewed without reference to the predator, has the appearance of competition between them, even if they are not actually competing for any limiting resources. Moreover, because individuals of each prey species suffer from the presence of the other, the evolution of different defense systems is favoured.

Most species are resources not only for predators, but for parasitic species of microbes, fungi, or animals. The ecology of parasites has been insufficiently studied, but they constitute a large fraction of the world's species, and have doubtless had a pronounced influence on the evolution of their hosts. Two of the most important aspects of the ecology of parasites are their reproductive capacity and dispersal. Whereas a single prey individual may be only one of many that a predator must eat before it can reproduce, a single host individual may yield an enormous number of parasite offspring. Thus the intrinsic rate of natural increase of a parasite is often many times greater than that of its host.

For many parasites, dispersal to new hosts is fraught with difficulty, and may be accomplished by any of a vast number of remarkable adaptations, such as infestation of a secondary host that is consumed by the primary host (in which the parasite reproduces sexually). The population density of the host often affects the probability of encounter, and therefore the growth rate of the parasite population. The genetic structure of a parasite population is often different from its ecological structure: the parasite population may be reproductively a single unit, yet ecologically be subdivided into groups of individuals that interact with each other within their individual hosts. Parasites are commonly quite highly host-specific, and in this sense are more specialized than most predators, which usually prey on a considerable variety of species.

Commensalism is a relationship in which one species benefits from another species that is not affected by the interaction; examples might include the dispersal of seeds on the fur of mammals, or the use of woodpecker holes by owls and chickadees. In many cases the evolution of one species may be affected by the other, but not vice versa. *Mutualism* is an interaction in which individuals of each species exploit the other as a resource, and so profit from the interaction. Examples include plants and pollinating animals, legumes and nitrogen-fixing bacteria, and the association between many vascular plants and fungi, which form a mycorrhiza, a root-fungus combination in which the fungus uses carbohydrates synthesized by the plant and the plant profits from a more rapid uptake of water and nutrients. Mutualisms may sometimes evolve from commensal relationships, but often seem to evolve from parasite-host relationships. Indeed, the line between antagonism and mutualism is often a fine one; the same fungus that enhances the growth of a plant in some environments may reduce it in others. Some mutualisms involve suites of more or less interchangeable species; thus most bees gather pollen from several or many plant species, and most bee-pollinated plants can be pollinated by several or many species of bees. Other, often highly obligate, mutualisms are highly species-specific; for example, leaf-cutter ants (*Atta*) feed exclusively on a particular fungus that they grow in their nests. The fungus is not found elsewhere.

Competition among species may take either of two general forms. Competition by *Exploitation* occurs when one individual consumes a resource and makes it unavailable to others; the competing individuals may never actually meet. Competition by *Interference* occurs when two individuals directly interact, and one loses the encounter. They may fight over food or territorial space; one individual may poison another (the phenomenon of allelopathy); or individuals may eat each other (which may be treated mathematically as a kind of competition). Competing individuals may be members of the same species (*Intraspecific competition*) or of different species (*Interspecific competition*). The most common model of interspecific competition is an extension of the logistic equation for population growth, in which the term $K - N$ is extended to $K_i - N_i - \alpha_{ij}N_j$:

$$dN_i/dt = r_{mi}N_i(K_i - N_i - \alpha_{ij}N_j)/K_i$$

This expresses the growth rate of species i (dN_i/dt) as a function of the relation between the carrying capacity for that species (K_i) and the number of individuals of both species i (N_i) and its competitor j (N_j); as N_i and N_j

increase, the term in parentheses decreases and dN_i/dt declines. The *Competition coefficient* α_{ij} expresses the impact cf one individual of species j on the growth rate of population i, relative to the impact of an additional individual of species i on its own population; α_{ij} equals 1 if the two species are equivalent in their competitive interaction, and is less than 1 if an individual of species j is a less serious competitor, from the viewpoint of population i, than a conspecific individual.

Species j might be a less serious competitor if, for example, it tends to use somewhat different resources, so that on average competition for resources is less intense between species than within species. Oversimplifying the result of this model, it may be said that species can coexist stably if competition within each species is more intense than between species (i.e., $\alpha_{ij} < 1$ and $\alpha_{ji} < 1$). If one species is a stronger competitor than the other (e.g., $\alpha_{ij} > 1$, $\alpha_{ji} < 1$), the inferior competitor will become extinct. If the species are competitively identical, their relative proportions will fluctuate by chance until one or the other becomes extinct. Thus this theory expresses "Gause's axiom," also known as the *Competitive exclusion principle*, that competing species cannot indefinitely coexist if they are limited by precisely the same resources. If they use only some of the same resources, they may be able to coexist. This principle is often invoked to explain the observation that coexisting species typically differ in the resources they utilize.

For example, several species of anoline lizards that coexist on West Indian islands forage in somewhat different microhabitats; species of triclad flatworms coexist only where the diversity of prey animals is high enough to provide each species with a food supply that it can largely call its own. Competitive exclusion is also a plausible explanation of cases in which species have complementary distributions. For example, the altitudinal distribution of the salamander *Plethodon glutinosus* is broader in mountain ranges where it is the only species than in some ranges in which *P. jordani* occupies the higher altitudes. Competitive exclusion has been historically documented in some cases; for example, within a few years after the parasite wasp Aphytis lingnanensis was introduced from Asia to control olive scale in California, it replaced the species *A. chrysomphali* that had flourished after its introduction some years earlier. Numerous studies have shown that competition exists among species of plants and of many animals in nature; when the density of one species is experimentally altered, the density of others changes in a compensatory fashion.

When two species compete, which one wins may depend on subtle environmental factors. Brown found that the chipmunk *Eutamias dorsalis* excludes *E. umbrinus* from low altitudes by chasing it away from food. But at higher altitudes the more arboreal *E. umbrinus* can get more food than *E. dorsalis* by quickly returning after being chased. It can do this because the branches of trees at these altitudes form an interlocking network, providing an effective highway that is not available in the sparser forests lower on the mountain. In interference competition of this kind, one species usually responds behaviourally to the presence or absence of another. For example, the lizards *Anolis sagrei* and *A. cristatellus* use different microhabitats where they coexist, but if *A. cristatellus* is removed, *A. sagrei* quickly expands into the vacant microhabitat. Some species, though, do not have this flexibility: the cricket *Allonemobius allardi* is replaced by *A. fasciatus* in moist habitats, but neither species expands its habitat range if the other is removed. As Darwin first realized, closely related species are likely to compete most intensely, but sometimes unrelated species compete as well. Ants and rodents compete for seeds;

terrestrial plant succession is a study in competitive exclusion of pioneer species by often unrelated late successional forms. Nevertheless, most studies of competition have focused on related species, often with the purpose of showing that they differ sufficiently in resource use to coexist. According to Gause's axiom, each species can persist only by differing somewhat from all others, so that a community should contain more species if each is specialized for a different resource than if the species are more generalized, and so overlap more in resource use.

A considerable number of studies have taken morphological differences among species as an index of differences in feeding habits, and have tried to show that there is a lower limit to the similarity of coexisting species. These efforts have been vigorously criticized and defended. In at least some cases, species that coexist do seem to be more dissimilar than if species had been assembled into communities at random. The Galapagos Islands, for example, harbour several species of ground finches with similar beak sizes and feeding habits, but these species do not coexist on any one island. One cannot attribute all differences in resource use to competition, nor will coexisting species necessarily differ in resource use.

Even species that do not compete can differ in feeding habits; for example, it is probable that many species of herbivores are limited by predation rather than resources, and there is little experimental evidence that they compete; yet related herbivorous insects often are specialized for feeding on different hosts. Predation on a superior competitor may prevent it from eliminating inferior competitors, and competitively inferior fugitive species may persist by colonizing new areas as they are eliminated by competition in others. Finally, more elaborate mathematical models than the one we considered show that it is sometimes theoretically possible for species to coexist stably even if they are limited by the same resource.

Beginning with Darwin, biologists have invoked competition as an important force affecting the diversification of species. On the one hand, competitors may prevent a species from evolving to use a particular resource; this is illustrated by its opposite, *Ecological release*, whereby a species uses a broader range of resources in the absence of a competitor. This may be reflected by morphology. For example, the males of many woodpeckers have larger beaks than the females and feed in somewhat different parts of trees. *Adaptive radiation* is a term used to describe diversification into different ecological niches by species derived from a common ancestor. Impressive examples include the Hawaiian honeycreepers and the cichlid fishes of the great lakes of Africa; in both groups, closely related species differ greatly in feeding habits and in associated morphology. As in many examples of adaptive radiation, these two groups have diversified in regions in which the diversity of other birds or fishes is low.

In theory, interspecific competition can favour the evolution of divergence so that species come to differ in the resources they use. Evolutionary responses of this kind may foster adaptive radiation, but they are not necessarily responsible for the adaptive radiations that have occurred. Evidence that species evolve in response to interspecific competition is provided by instances of *Character displacement*, defined as a greater difference between *Sympatric* (in the same geographic locality) than *Allopatric* (in different localities) populations of two species. A few examples of this phenomenon are known, but not very many.

We are accustomed to thinking in simplistic, linear terms: the more abundant mice are, the more abundant the weasels that feed on them. But interactions among species are seldom so simple. Predation can sometimes prevent a superior competitor from excluding an inferior one. The intensity of competition between two species can be modified by the presence of a third competitor.

Two consumers may enhance each other's population growth by preventing competitive exclusion among the species they feed on. A species may have either a net beneficial or a net deleterious effect on another, depending on the presence or absence of a third species. For example, oropendolas (large tropical American orioles, *Zarhynchus wagleri*) are often aggressive toward giant cowbirds (*Scaphidura oryzivora*) which lay eggs in oropendola nests. This response is not surprising, because young cowbirds compete for food with young oropendolas. But young cowbirds can enhance the survival of oropendola nestlings by picking botfly larvae off them. However, oropendola nests in trees that harbour large wasp nests are rather free from botflies because wasps chase botflies away. Remarkably, the oropendolas are hostile to cowbirds when they nest in the vicinity of wasps, but tolerant of cowbirds in the absence of wasps. Even more remarkably, cowbirds that lay in oropendola nests associated with wasps lay mimetic eggs, while those that parasitize oropendolas that are not associated with wasps tend to lay non-mimetic eggs. Some species have a profound impact on the distribution and population dynamics of others simply through their physical effect on the environment.

For example, the periwinkle (*Littorina littorea*), an herbivorous snail introduced to the American coast from Europe, acts like a little bulldozer when it grazes. When periwinkles are removed, the community of barnacles and encrusting algae is replaced by a community of filamentous algae and benthic worms and snails that live in the sediment that accumulates. Finally, we should bear in mind that many species subsist on the *interaction* between other species. Numerous parasites, for example, depend on predator-prey relationships, as in some trematodes that develop sequentially, first in snails, then in snail-eating fishes, and finally in piscivorous birds. The development of species diversity in a community is self-reinforcing, as interactions among species create new ecological niches.

Ecologists have devoted a great deal of attention to the question of what deter mines the number of species (species diversity) in communities, and whether there exist forces that organize communities into a predictable structure. The number of species—of birds, for example—often is related to the area embraced, more or less according to the relation $S = cA^z$, where S is the number of species, A is the area, and c and z are constants; z often has a value of about 1.2. The theory of island biogeography explains the relationship by noting that the number of species in a region will be set by the balance between the rate of extinction and the rate of immigration of new species into the area. There is some evidence that extinction rates are higher in small areas, which support smaller populations, than in large areas. It is unlikely that communities contain a random assortment of species that have chanced to invade them; rather, it appears that interactions among species permit some combinations to persist while excluding species that do not fit in. One reason for the belief that communities are organized is that faunas that have independently adapted to similar environments in different parts of the world sometimes appear convergently similar—although this is not always true. Moreover, as we have seen, related species in a community sometimes appear to be less similar in their use of resources than if the communities had been assembled by chance. Although there is considerable controversy about the interpretation of such data, few ecologists would altogether deny that some combinations of species are stable, whereas others are unstable, in the sense that one or more of the species will be unable to invade the community, or will become extinct if another invades.

Any physical or biotic feature of the environment that impinges on individuals of a species has several different properties which may affect the course of adaptation. For instance, there is its average value. We commonly presume that some average conditions are harsher, or more difficult to adapt to, than others. This may well be true, but it is hard to measure harshness, because it is relative to the capabilities of the particular species.

A warm rainforest would presumably be as harsh for a caribou as the tundra would be for a spider monkey. Ephydrid fly larvae that inhabit pools of crude oil seem to do quite well, although this environment is uncongenial for the vast majority of species. Adaptation to a "harsh" environment is more likely for some species than others, and one of the major challenges to evolutionary biology lies in understanding why some adaptive changes are more likely than others. Virtually all environmental factors vary. It is important to distinguish *spatial* variation from *temporal* variation. The factors that affect white-tailed deer (*Dama virginiana*) in Maine do not appreciably affect the survival of white-tailed deer in Florida. Local populations, exposed to different states of the environment, may take quite different evolutionary paths.

However, the spatial scale over which the environment varies effectively depends on the distance over which members of the species typically move. Thus a single genetic population of the migratory monarch butterfly (*Danaus plexippus*) experiences the temperature regimes of both Canada and Mexico, whereas different populations of a nonmigratory species experience very different temperature regimes at different latitudes. Temporal variation in the environment may be described in several ways.

The *Absolute range* of the variable includes extremes that may be quite rare; a really cold winter once a century or an outbreak of a defoliating insect that ordinarily poses little threat to a tree species may have a dramatic impact on a population. Such unusual events must be part of the evolutionary experience of all species, but we can seldom assess their impact on evolution, even in theoretical terms.

An important aspect of environmental variation is its frequency. The effect of the frequency of oscillation depends on the "response time" of the physiological function that is affected by the environmental change. For example, because of its large size a lion has a relatively low metabolic rate and can store considerable fat, so it requires a large meal only every few days; but if food is available that infrequently to a shrew, it will quickly die because of its far higher rate of metabolism. Thus an environment that is effectively constant for a large homeotherm is lethally inconstant for a small homeotherm. It is often useful to distinguish between environmental fluctuations with a period less than the life span of an individual organism (*Fine-grained variation*) and those with a period greater than an individual's life span (*Coarse-grained variation*). Fine-grained variation is often met with homeostatic changes in physiology or behaviour that are reversible. Adaptation to coarse-grained variation often entails a developmental switch into one of several irreversible alternative phenotypes. For example, if the rotifer *Brachionus* is exposed early in life to the chemicals that exude from the predatory rotifer *Asplanchna*, it develops long protective spines. In many semiaquatic plants the shape of the mature leaves depends on whether or not they are submerged during their development.

The distinction between fine-grained and coarse-grained environmental variation holds for both spatial and temporal heterogeneity. The variety of plant species in a forest is a fine-grained mosaic to a leaf-eating monkey that in its lifetime can sample all the available plants. But to an individual gall wasp that undergoes its entire development in a single leaf, the forest is a coarse-grained environment.

The consequences of environmental variation depend in part on the time scale over which a variable fluctuates. Some variation exists on a physiological time scale, within the lifetime of an individual organism, whether because the individual moves about from one environment to another or because it is buffeted by change even if it stays in place.

A major consequence of such variation is to select for genotypes that have homeostatic abilities or that can avoid exposure to inimical changes (e.g., by dormancy). There exists variation also on an ecological time scale of greater than a generation—changes in physical or biotic variables that bring about changes in population density, including local extinction. Ecological succession is just one example of such change: early successional species such as dandelions become excluded locally by late-successional species, and persist only as other patches of favourable habitat become open elsewhere. As long as suitable habitats become available at a fairly constant rate, however, such *populations* may be said to experience a fairly constant environment, even if many of the *individuals* that compose these populations experience a deteriorating environment. As we consider longer time spans, variation on the ecological time scale grades into variation on a geological time scale. A new environment that persists for several to many generations will often bring about changes in the genetic composition of the population by natural selection, as well as changes in population density.